Sundials

Armillary-type Dial in Hamburg, Germany.

Sundials
THEIR CONSTRUCTION AND USE

THIRD EDITION

R. Newton Mayall and Margaret W. Mayall

SKY PUBLISHING CORPORATION
CAMBRIDGE, MASSACHUSETTS

Third Edition, 1994

Published by:
Sky Publishing Corp.
49 Bay State Rd.
Cambridge, MA 02138-1200

ISBN 0-933346-71-9
LC Catalog Card Number: 73-76242
Printed in the United States of America

94 95 96 97 98 8 7 6 5 4 3 2

TO
OUR
PARENTS

PREFACE TO SECOND EDITION

W HEN this book was first published there was no easily obtainable information on how to make various kinds of sundials. Our intent was to make the basic know-how available to anyone with an interest in dialling.

Since then we have found a desire among our readers for more information. Many have acquired their own collection of portable dials, and you too, with some imagination, can make literally hundreds of different designs. Yet the basic principles remain the same.

In this edition we have made only a few changes in the original text but have added chapters on constructing portable dials, dials with variable centers (the analemmatic type), and a simple heliochronometer. Another chapter is devoted to a classification for cataloging sundials.

It is hoped that these additions will engender an even greater interest in dialling.

R. NEWTON MAYALL
MARGARET W. MAYALL

PREFACE

$\mathbb{A}$ PORTION of the material in this book first appeared as a series of articles in the Scientific American. The purpose of the articles was to show the construction of sundials, their use as accurate timekeepers, prove that they are not just garden ornaments, and make available material not easily obtained. The response to the series indicated a need for a comprehensive book dealing with the construction and use of the sundial. So much sentiment has been woven about this instrument of antiquity, its utility has apparently been forgotten.

A good sundial will show the time of day as accurately as many watches; and Standard Time, which is in universal use today, may be easily obtained from it.

To most people the construction of a sundial implies laborious mathematical calculations and a knowledge of astronomy. Such, however, is not the case. No words of ours could better express the purpose and content of this book than the statement which appears in the preface of Leadbetter's treatise on dialling, published about the middle of the 18th century—"Dialling, if mechanically considered, is of itself a thing so natural and easy, one would wonder, after so much learned bustle as the mathematicians have made about it, that they should have more perplexed and obscured than promoted the knowledge of that useful and entertaining art amongst the generality of mankind."

The major portion of this work describes the construction of the hour lines for many kinds of sundials, by the graphic or geometric method. The use of this method does not require a knowledge of mathematics or astronomy. Its simplicity and accuracy, together with the ease and quickness of delineation make it very practical. Only common materials available in the average household are needed—such as, paper, pencil, straight-edge, compasses and protractor.

Sundials may be described on almost any surface, in any position. Rarely, however, does occasion arise for constructing them on any but a plane surface, and in either a horizontal, vertical, or reclining position. There are many ways of laying out the hour lines on each type of dial; only the most accurate constructions have been illustrated, and all of them have been carefully checked by mathematical computation.

In order to fill the requirements of all, the text has been further augmented by the addition of a chapter on formulas, wherein is set forth the trigonometrical formulas for computing the hour lines for various types of dials, with examples of the computation where necessary.

Many variations of the dials described can be made, and all of them may be adapted to portable use. The construction of portable dials is a fascinating pastime. There are several collections of portable dials in the United States, but many of them are private. Among the most noteworthy collections, open to the public, are those in the Adler Planetarium, Chicago; Columbia University, New York City; the Metropolitan Museum of Art in New York City; and the Harvard College Observatory, Cambridge, Massachusetts.

It is our hope that this volume will not only serve as a practical handbook, but also engender a further interest in sundials. The illustrations have been selected with care, many of them published here for the first time. No effort has been made to include material that is easily obtained elsewhere— for instance, if one wishes to see pictures of portable dials, there are several good books available containing many photographs and drawings of them, such as "Sundials and Roses of Yesterday" by Alice Morse Earl and "The Book of Sundials" by Mrs. Alfred Gatty. These two books are also excellent references for mottoes.

We are grateful to the editors of the Scientific American, who first brought much of this material to the attention of the public, for their cooperation and courtesy in extending the use of many plates and illustrations reproduced here. Hundreds of queries by readers of the Scientific American, from all parts of the world, have been our constant guide in the preparation of the contents.

Acknowledgments are due to Dr. Harlow Shapley, Director of the Harvard College Observatory, for many courtesies, his interest, and much constructive criticism; also to Dr. Loring B. Andrews of the Harvard College Observatory, whose interest has been a great encouragement.

To all others who in any way have aided in this work— our thanks.

R. NEWTON MAYALL
MARGARET L. (WALTON) MAYALL

CONTENTS

I

THE DEVELOPMENT OF THE SUNDIAL

IT IS not at all surprising that the present generation knows little about the sundial, which in our present complex existence has become the forgotten timekeeper. It is reminiscent of a more leisurely existence when "time waited for no man", whereas today no man waits for time.

When the Pilgrims landed on our shores and up to the time of the American Revolution sundials were the most common timekeepers on the Continent, even though many cities and towns had erected towers containing primitive mechanical clocks similar to the one in Milan, Italy; and despite the fact that at the beginning of the 20th century mechanical timekeepers had been perfected, sundials were still used by one of the leading railroads in France to regulate the watches of their trainmen. Furthermore, how many, except perhaps the most adventurous travelers, know that in many places throughout the world the sundial is, even today, the principal or only timekeeper; that in parts of Japan and China, a simple noon mark dial is used by government post offices. A recent letter from a postmaster in a small Japanese country town states that he uses a noon mark dial "to regulate the time and it is quite punctual than to depend on cheap watches."

Man has always regulated his life and work by time in one form or another. Primitive man may have been content with a day of two periods—starlight and sunlight. As it became necessary for him to travel farther afield he soon would have observed that a constant watch must be kept on the apparent motion of the sun in the sky. He could travel outward as long as the sun rose, but as it began its descent toward the opposite horizon he must hasten to retrace his steps in order to return before nightfall. This division of the day into two parts must soon have become insufficient. It is not improbable that primitive woman may have caused man to devise a means of apportioning the day into smaller parts which could be relied upon, for reasons easily imagined. His solution to the problem is readily conjectured. Surely our caveman ancestor noticed the phenomena of shadows cast by upright objects—how the shadows lengthened and shortened in relation to the position of the sun. By placing a stick firmly in the ground he could watch and study the shadow it cast. Probably stones or sticks were placed at the extremity of the shadow at various times during the day, giving him definite periods of short duration, and the length of these periods could be arranged to suit his comfort and needs.

But, another problem arose. How could the traveler return at a prearranged time? Here again the solution is obvious to us. He could carry with him a stick equal in length to the height of the one which had been securely placed in the ground near his cave. Thus, the first stationary and portable sundials may have been born. No doubt Mrs. Caveman frequently remarked, "Don't forget your shadow pole and return when the shadow's length is one pole."

If the perpendicular stick or gnomon type was the first dial, there is nothing to indicate what was produced between the time of its invention and the appearance of those early dials of which we have information. In order to preserve clarity and continuity in tracing the development of the dial, no detailed definitions of terms explained elsewhere will be given here.

Therefore we leave conjecture behind and let the sundial

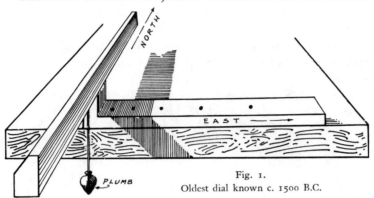

Fig. 1.
Oldest dial known c. 1500 B.C.

tell its own story, beginning about 1500 B.C. At the beginning of the 20th century the earliest dial known was devised about 370 B.C., whereas today we have examples of dials used in Egypt about 1500 B.C., which were brought to light through archeological exploration. As the archeologist has made us more familiar with the life and work of early peoples, so has our knowledge of early timekeeping instruments penetrated the dark recesses of history.

We know the Egyptians were well versed in astronomy and mathematics; that they understood at a very early date the motions of the earth and planets; and that they had fixed the year at about 365 days; but, very few Egyptian sundials have been found. However, the oldest dial, Figure 1, is among them. This dial was made of stone in the form of a

flat bar about 12 inches long with a perpendicular T-shaped
piece fixed at one end. The time of day was deduced by the
position of the shadow cast, by the upper edge of the cross
piece, between the marks cut at irregular intervals on the top
surface of the bar. When in use, the cross piece must be
turned toward the east in the morning and toward the west
in the afternoon. The plumb line is used for placing the instrument in a level position.

Figure 2 shows another Egyptian dial of similar character
constructed during the period of about 660-330 B.C. (Later

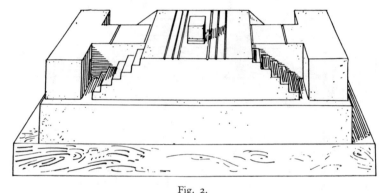

Fig. 2.

Egyptian dial c. 660–330 B C.

Period), which tells time throughout the day without being
turned for the afternoon hours. In addition to the flat dial
surface, ramps and steps have been cut into the sides. The
position of the shadow on them will also give the hour. This
arrangement enabled the dial to be set without the aid of a
standard line or meridian, for it was only necessary to place
it in a level position, then move it until the time shown by
the shadow on the ramps or steps agreed with the time shown

on its upper level surface. Such a dial if made small and light enough, could easily be carried about.

One more Egyptian dial, Figure 3, is of particular interest. It is of the period about 330-30 B.C. (Hellenistic Period), and shows a decided advance over the previous dials in that in-

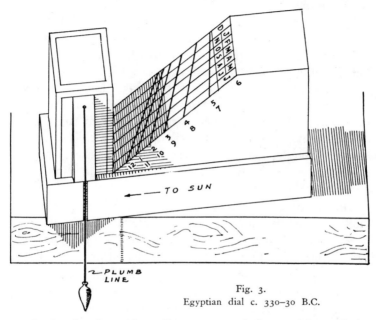

Fig. 3.
Egyptian dial c. 330–30 B.C.

stead of a horizontal surface to record the position of the shadow cast by the upper edge of a perpendicular block or gnomon, the surface was inclined at an angle equal to the latitude of the place. Its width was divided to show the months and across these divisions diagonal lines were drawn representing the hours of the day. When in use the instrument was first placed in a level position by means of the plumb line, then turned so that the perpendicular block was pointed directly toward the sun. The position of the shadow upon the hour lines corresponding to the proper month would show the time for any day. This was an ingenious device, because the Egyptians did not make use of hours of equal length, as we do today—they used temporary or unequal hours.

Temporary hours resulted from the division of the period between sunrise and sunset into twelve equal parts. Because the length of this period varies throughout the year, it was not possible to obtain equal divisions of time by such a method except on any one specified day. Therefore it was necessary to observe the position of the shadow at each hour on several days during the year, preferably at the time of the equinoxes and the summer and winter solstices. If lines drawn through these points were crossed by others designating the months the true temporary time could thus be obtained any day in the year.

Timekeeping was not the only incentive for making these dials, for they were often used as votive offerings and placed in temples. The period of production is our only clue to the age of Egyptian dials—their makers are unknown. A contemporary device—the clepsydra or water clock—made it possible to tell time at night or when the sun did not shine, by measuring or indicating the height of water in some receptacle from which the flow could be regulated.

We must now retrace our steps a few centuries to pick up the threads of a lost sequence. Those who are familiar with their bibles will remember Ahaz was the King of Judah about 742–727 B.C. Perhaps you will even recall the "Dial of Ahaz", attributed to one of his Babylonian astronomers, which is mentioned twice in the scriptures:— In II Kings XX:9–11

"And Isaiah the prophet cried unto the Lord; and he brought the shadow ten degrees backward, by which it had gone down in the dial of Ahaz."

and in Isaiah XXXVIII:8

"Behold, I will bring again the shadow of the degrees, which is gone down in the sundial of Ahaz, ten degrees backward. So the sun returned ten degrees, by which it had gone down."

This phenomenal movement of the shadow on the Dial of Ahaz has given rise to as much discussion as the squaring of, the circle and the trisecting of any angle. For years, it has puzzled layman and scientist alike. The form of the dial remains a matter of conjecture.

More than a century after the reign of Ahaz we learn of a dial erected, about 560 B.C., by Anaximander of Miletus (611–547 B.C.), a Grecian astronomer. This was probably a vertical rod or gnomon erected in the public square, similar to, but more carefully constructed than the upright stick of the caveman, because more information about the movement of celestial bodies was at hand as evidenced by the work of the Egyptians.

The Chaldeans had made substantial progress in mathematics and astronomy. By constant observation of the heavens they became familiar with the constellations and saw in them the likenesses of human beings and animals; they divided that band in the sky called the "Zodiac", in which the sun and planets move, into twelve parts or signs each containing a configuration, named and referred to as the Zodiacal constellations. They also divided the year into twelve parts, devised the week of seven days, and foretold eclipses.

One of the simplest forms of the sundial—the hemispherium, Figure 4—is attributed to the Chaldean priest and as-

tronomer, Berosus, who lived at the time of Alexander the
Great (356–323 B.C.). This dial was carved out of a block
of stone, its concave hemisphere resembling the inverted
vault of the heavens. A perpendicular pin or style was placed
in the center, pointing to the zenith; then as the sun traversed

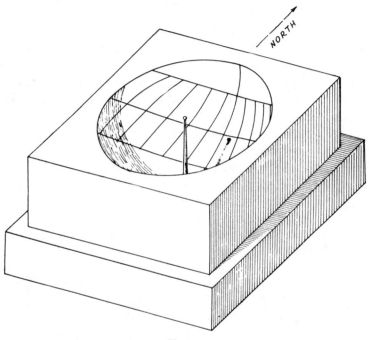

Fig. 4.

the sky, the shadow of the top of the pin would trace out the
apparent motion of the sun in a reverse direction. That portion of the inner surface upon which the sun shone was divided into twelve parts representing the temporary hours.
The hour lines were crossed by three or seven other lines cor-

responding to the seasons or months, which were determined by the same method used in Egypt.

Although inaccurate, the hemispherium was far superior to the waterclocks in common use at the same time, because they were bulky, needed attention and could not be carried about easily; whereas the hemispherium could be made small enough to be carried in the pocket and set up anywhere.

The hemicyclium, Figure 5, is also attributed to Berosus and it is often referred to as the "dial of Berosus." There is a difference of opinion concerning the inventor of these two dials which may be due to a loose use of the two words in modern literature as meaning the same kind of dial. Al-

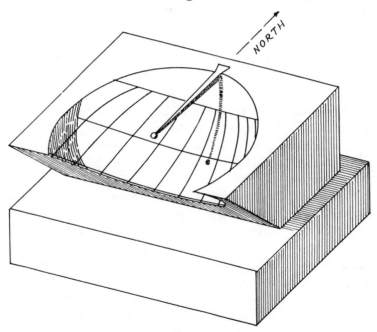

Fig. 5.

though there is no difference in the construction of the lines, the hemicyclium has the front or south portion cut away at an angle, and a horizontal gnomon is used instead of a perpendicular one. The portion cut away is useless for timekeeping purposes because the shadow would never enter that part. Some early writers considered the hemicyclium a great improvement over the hemispherium, which was probably due to the greater ease in reading, and its lighter weight. Both dials were made in forms and sizes too numerous to mention.

Although the introduction of Euclid's "Elements" (ca. 300 B.C.), with which all of us have struggled at one time or another, gave great impetus to the progress of mathematics, no great improvement was made over the hemicyclium for many years. The writings of Albategni show that these concave dials were commonly used in Arabia as late as 900 A.D., and the same construction was followed.

About 100 years after the appearance of Euclid's work, Apollonius of Perga (250–220 B.C.) made public his treatise on the theory of conic sections, which laid the foundation for the geometry of position. The advent of this new study soon brought about a change in sundials, resulting in the conical dial, Figure 6.

The conical dial was an improvement over previous dials in that its essential factor was greater accuracy. Its appearance was not unlike that of the hemicyclium, although the concave segment of a circular cone was used instead of the hollow section of a sphere. The surface was delineated in much the same manner, with the twelve unequal hour divisions crossed by three or seven arcs corresponding to the seasons or months. Very few dials of this type have been found in the ruins of Egypt, Greece, and Italy; but they were probably not introduced before 200 B.C. Either the lower surface or axis of the dial was inclined at such an angle that it pointed to the north star. At this time a wider knowledge of conic sections was necessary to further improve the sundial.

After the fall of Alexander the Great we find such names as Aristarchus (ca. 280–264 B.C.), Hipparchus (160–125 B.C.), and Strabo (29 B.C.–14 A.D.). Hipparchus was the founder of scientific astronomy and it was he who laid the foundation for our present trigonometry. Contemporary scientists quickly grasped this new method of computing, but it was left to others, later, to apply the theory to the improvement of the sundial.

We now find ourselves at the beginning of the Christian

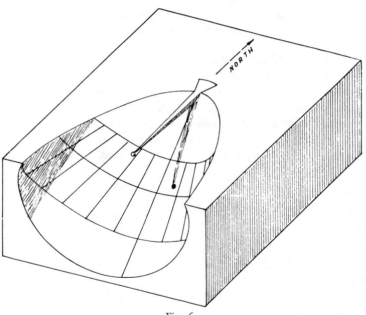

Fig. 6.

Era, with a picture cf the type of dials evolved by the Egyptians and Greeks. Through the centuries little advance had been made in timekeeping quality.

Other countries employed the sun as a timekeeper—the Arabians attached great importance to the science of sundial construction, which they learned from the Greeks. They had dials of similar construction, as did the Romans who also adopted them from the Greeks. They did not add any new types or evolve new methods of figuring and laying out the hour lines. That dials were prevalent in Rome and throughout the Empire is evident by this choice morsel from the pen of Maccius Plautus (ca. 250–184 B.C.), a comic poet and writer of that city:

> "The gods confound the man who first found out
> How to distinguish hours! Confound him, too,
> Who in this place set up a sun-dial,
> To cut and hack my days so wretchedly
> Into small portions. When I was a boy,
> My belly was my sun-dial; one more sure,
> Truer, and more exact than any of them.
> This Dial told me when 'twas proper time
> To go to dinner, when I had aught to eat.
> But now-a-days, why, even when I have,
> I can't fall-to, unless the sun give leave.
> The town's so full of these confounded dials,
> The greatest part of its inhabitants,
> Shrunk up with hunger, creep along the streets."

Be that as it may, we are indebted to the Romans for a valuable contribution, which is found in the "Treatise on Architecture", written by their renowned architect Vitruvius, who died during the reign of Augustus (30–14 B.C.). This is the only accessible literary work of Roman origin mentioning sundials. We cannot lightly skip over this ancient record, for Vitruvius says that he will "state by whom the different classes and designs of dials have been invented. For I cannot invent new kinds myself at this late day, nor do I think that I ought to display the inventions of others as my own." He lists thirteen dials, in the following order:

"The HEMICYCLIUM of Berosus
The HEMISPHERIUM of Aristarchus
The DISCUS ON A PLANE of Aristarchus
The ARACHNE of Eudoxus
The PLINTHIUM of Scopas
The UNIVERSAL DIAL of Parmenio
The UNIVERSAL DIAL of Theodosius and Andrias
The PELICONON of Patrocles
The CONE of Dionysidorus
The QUIVER of Apollonius

"The men whose names are written above, as well as many others, have invented and left us other kinds: as for instance, the CONARACHNE, the CONICAL PLINTHIUM, and the ANTIBOREAN."

The appearance of a few of the dials listed is known, but there is no definite knowledge about the rest. Vitruvius credits Aristarchus with the invention of the hemispherium. Presumably he has arranged the list of dials in order of their age (as is the sequence of diagrams included in this chapter), but it would have been more logical to place the hemispherium first because of its simplicity. Furthermore Berosus preceded Aristarchus by 100 years.

Vitruvius also mentions the fact that other writers have left directions for the construction of dials for travelers, "which can be hung up." He then states that anyone can construct such dials from the directions in books on the subject, "pro-

vided only he understands the figure of the analemma."
(The analemma was an instrument and method of projection
which demonstrated and solved some of the common astronomical problems,—not the figure eight we see on modern
globes and atlases. It is not of sufficient importance here to
warrant detailed explanation, which can be obtained from
any modern reference work.) This leads to the supposition
that, up to the time of Vitruvius, no geometrical method
had been evolved to construct the hour lines for a dial.

At the end of the pre-Christian period the use of sundials
extended over the greater part of the Western World, but up
to that time few improvements had been made on the known
types other than simplifying their form and the addition of
certain embellishments.

One of the most interesting monuments of the 1st century
B.C. is still in existence—the Tower of Winds at Athens,
Greece. This octagonal tower has not only been admired by
students, but the dials on its eight faces have provoked much
discussion as to their origin and date. Vitruvius mentions the
tower but does not refer to the dials. This is the argument
used by those who believe them to have been added at a later
date. No one has proved whether or not they were a part of
the original structure, and many interesting papers have been
published on both sides of the question.

Further development was not accomplished until after
the beginning of the Christian Era when it was discovered
that by slanting the gnomon so that it pointed exactly to the
celestial pole, Figure 7, the direction of its shadow could be
made to show solar time correctly; for the gnomon would
then lie parallel to the earth's polar axis and the sun, moving
parallel to the celestial equator, would always move straight
across the gnomon or appear to revolve about its sloping edge.

The apparent east to west motion of the sun alone governed the swing of the shadow and the dial would keep true time with the sun every day in the year. Nothing remained to be done, except to derive the necessary formulas for computing the true direction of the shadow for each hour of the day, on any surface.

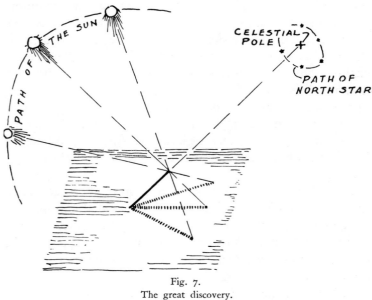

Fig. 7.
The great discovery.

Just when or by whom this great discovery was made or the dial so scientifically perfected is not known, but it probably occurred in the 1st century A.D. That it could not have occurred before the time of the Later Greeks is quite certain, although the Chaldeans and Egyptians were quite capable of understanding this principle, which was the natural result of a more scientific study of the heavenly bodies. Simple as it may seem today, it soon revolutionized dialling and many

changes in form were made. The application of trigonometry to this principle opened wide the door to more accurate
timekeeping by means of the sun.

In the 2nd century of the Christian Era, Ptolemy (139–
161 A.D.) wrote that great treatise on mathematics and astronomy called the "Almagest", wherein are given instructions for constructing sundials by the use of the analemma
which enabled one to project the direction of a shadow geo-
metrically. This was a departure from the method used by
the ancients who marked the hours by the end of the shadow
of the gnomon. Ptolemy shows the construction of the hour
lines on different surfaces, in a variety of positions.

Although the analemma made it possible to construct dials
more easily and accurately than before, two things retarded
further perfection—the use of hours of equal length had not
been generally accepted, and trigonometry required further
development.

The fall of the Western Roman Empire about 400 A.D.
further set back the development of the sundial and marked
the beginning of that period of intellectual obscurity known
as the Middle or Dark Ages, during which time practically
no important advance was made in any of the sciences with
the exception of mathematics.

The development of trigonometry is attributed to Albategni (850–929 A.D.), an Arabian, whose work has been preserved. It made possible the precise calculation of sundials, and the application of the new trigonometry to their
construction led directly to the final step in the development
of the dial. From that time on nothing was impossible as
was evidenced in the next period of importance—the Renaissance.

As the world emerged from the period of intellectual dark-

ROTHENBURG, GERMANY

Near the top of this building on the north side of Market Square in Rothenburg, Germany, is a 3'-by-4' dial dated 1768. Below it are two other indicators, which give both the date and the zone time. Note the discrepancy between the time indicated on the dial (12:35) and the face of the clock (12:52).

ness, the sundial had reached a certain state of perfection. Abul-Hassan, an Arabian who lived about the beginning of the 13th century is said to have introduced equal hours, such as we use today. Evidently his invention was not well received, for it is not until about the 15th century that we find their general use well established.

The transformation from the Dark Ages of the 5th to 13th centuries to the Revival or Renaissance Period (14–16th centuries) marked the transition from the ancient to the modern sciences of astronomy and mathematics. Copernicus (1473–1543) upset the picture of our universe as presented by his predecessors. His theory that the sun and not the earth was the center of the solar system slowly but surely revolutionized the science of astronomy. The acceptance of this theory and equal hours, together with the attendant application of trigonometry, gave us our present scientific sundial.

During the Renaissance sundials were made in every conceivable form, on all possible surfaces, and in every possible position or location.

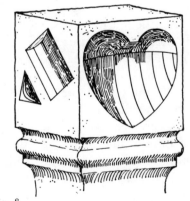

Fig. 8.
Sunk dials

The so-called "sunk dial", Figure 8, was outstanding in this period. It became so varied in form that it bore little resemblance to its venerable predecessors—the hemicyclium and hemispherium. They were heart shaped, cylindrical, triangular, and so forth. Not content with one dial, sometimes one stone would have as many types as it was possible to crowd on. Such dials were placed on headstones in cemeteries, on the buttresses on churches, and others were monumental in character their real purpose often hidden by their grotesque appearance. The sundial became a proper gift for kings and queens, and the use of both portable and stationary forms spread rapidly throughout Europe. Portable dials were made in as many varieties as the stationary dials. They were small enough to put in the pocket and many of them were adjustable to suit various localities, others were fitted with compasses to aid in setting them; and still others were adapted to use at sea by mounting them on gimbals like chronometers. Dial makers vied with each other for supremacy, kept their methods secret, shrouded them in mystery, and construction became a lucrative occupation. There was little left to be desired in the way of sundials and many fine examples of this period may be seen today in Scotland. The Renaissance marked the culmination of timekeeping by the sun.

We must leave the ancestral home of the dial in the Euphrates valley and Egypt in order to pursue its development as its influence stretched out toward the northern countries of Europe and the British Isles; for it is in these countries we must look for advancement in astronomy.

Travelling northward through Italy, France and Germany to England we encounter the names of many famous astronomers and mathematicians, who made outstanding contribu-

THE DEVELOPMENT OF THE SUNDIAL

tions to these two sciences; among them are Galileo, who said
of the earth "but still it turns"; Kepler, whose laws of planetary motion are still good; Newton, who observed the apple
fall to earth; Herschel, a great observer; and Delambre, scientific historian whose record of early sundials and history of
astronomy is invaluable.

These men lived in the period 1500 to 1800 A.D., a period
significant in the development of the sundial. Men well
versed in mathematics and astronomy seem to have felt it
their duty to acquaint everyone with the theory of constructing dials. It was all so simple that even the uneducated peas-
ant should know how to build his own sundial. The reason
for this change in thought from secretiveness to openness
and the tremendous influx of literature was due to the invention of moveable type for printing, without which the production of a book was a laborious process.

The writers of this period did more to make the sundial
of practical value than did those of any other period in history. They simplified formulas and devised many methods
by which the layman could lay out the hour lines. The art of
designing and constructing dials accurately was no longer
confined to the craftsman, mathematician, and astronomer.
The dial became a scientific instrument, more dependable
and lasting than any mechanical device, its only disadvantage
that the sun must shine. They were also constructed so that
the time could be told at night either by observing the stars
or by the position of the shadow cast when the moon shone.

In addition to producing a timekeeper, they showed how
it was possible, by the addition of various lines called "furniture", to obtain the day of the year, the height of the sun, the
time of sunrise and sunset; the azimuth of the sun, and even
feast days were recorded by the shadow. Presumably one's

wedding anniversary may also have been marked to remind forgetful husbands.

It would seem that at the beginning of the 18th century the final chapter had been written, for watches were coming into favor; but, strange as it may seem, the invention of the watch really advanced rather than retarded the use of sundials. Most of the mechanical clocks and watches of this period were none too accurate and sundials were often used for checking and setting them. Shakespeare alludes to this fact in Scene I Act 3 of Love's Labour's Lost, where Biron says—

> . . . . "I seek a wife!
> A woman, that is like a German clock,
> Still a-repairing; ever out of frame;
> And never going aright, being a watch,
> But being watch'd that it may still go right!"

Mechanical timekeeping offered competition to the dial makers. If sundials were to maintain a position of practicability it would be necessary to have them show the same time as that shown by the clock.

The clock assumes that each day is of the same length and divided into twenty four equal parts or hours; whereas the sundial, although marking off equal hours, shows apparent or true solar time. Furthermore the sun's position in the sky varies from day to day. Obviously, therefore, it would not be practical if not impossible, to make a mechanical clock that would follow the vagaries of the sun. A difference, called the equation of time, was observed between clock time and sun time. This difference was found to be variable through the year, therefore it was possible to publish tables or charts showing the variation for each day in the year. Such tables and charts often accompanied dials, thereby making it possible to obtain clock time from the sundial by simple addition or subtraction. Other dials were so constructed that it was possible to tell clock time by a direct reading of the dial. In this manner, dials held their own.

By the early part of the 19th century the sundial began to lose ground as a timekeeper of importance. Man became more exacting in the use of time, not only in the sciences, but in his daily life. It was inevitable that the watch should be the instrument to suppress the sundial, because it embodied everything necessary for a good timekeeper—it showed the time of day rain or shine; it was small; it was portable and easily carried on the person; and it showed the time in any locality.

The refinements and improvements made in watches did not completely obscure the sundial but rather made it less necessary. The march of progress however brought about a change in timekeeping which again brought the sundial out into the light. This was the necessary adoption of a prime meridian from which all longitude should be reckoned. Such a meridian was adopted in 1884—the meridian of Greenwich, England—which was followed by the adoption of standard time zones, which are generally in use today throughout the world.

The adoption of standard time zones gave the dial makers another problem to overcome, if they were to justify its use as a timekeeper. Early watches were none too reliable, but when standard time was put into effect, they had been perfected, and it was no longer necessary to carry two for fear one might go awry. It might seem that by the middle of the 19th century the sundial had reached the acme of perfection and accuracy, but the acceptance of Standard Time urged the dial makers on to new heights. Many of these dials were scientific instruments, as beautifully machined and accurate as a surveyor's transit. They may be properly referred to as "20th Century Dials".

One type of dial has not been mentioned—that which makes use of a ray or beam of light passing through a narrow slit or small hole, often referred to as a "spot" dial, which was quite numerous between 1500 and 1800, particularly in the portable form. The use of light instead of a shadow for telling time was much more to be desired because the shadow was indistinct when cast on some metals; whereas

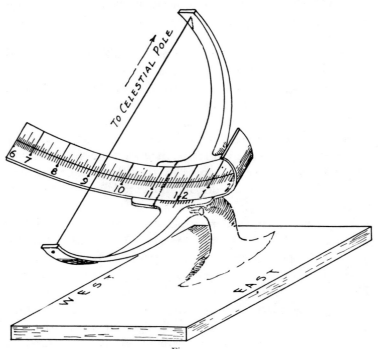

Fig. 9.
Simple heliochronometer.

the spot or beam of light stood out prominently, due to the shade about it. This device was often used on the new dials of great accuracy, such as the heliochronometer, Figure 9. Mirrors and lenses were also employed to direct the light.

The heliochronometer or sun clock is, as the name suggests, the peer of all sundials, usually quite heavy and beautifully machined. Fine wire is generally used to cast a shadow; a narrow slit or hole to direct light. It is often fitted with gears and a scale for subdividing the units on the face into parts of a minute, or even into seconds, of time. The base may be fitted with levels and adjustment screws; and above all the modern heliochronometer tells Standard Time. This is the type of dial used by the railroads in France, for setting watches, as late as 1900. If more dials of this type were constructed today there would be a still greater interest in sundials. The accuracy of such instruments when properly constructed and set up would probably amaze most people.

Sundials were used elsewhere than in European and Mediterranean countries, but their development and refinement was confined to those countries. No worthy contributions were made by other parts of the world. The Japanese and Chinese still use portable dials, which are copies of those developed in the Occident. It is strange that the hemispherium is one of the commonest forms used in the Orient—one of the oldest dials known, still serving a useful purpose. Of course other dials are used, such as the tablet form, Figure 10, which can be quite easily obtained today. It consists of two hinged wood blocks or tablets, which are placed perpendicular one to the other when in use and folded when carried. Upon one tablet is a vertical dial which faces the south; on the other is a horizontal dial. The shadow is cast by a string stretched between the two. A variation consists of adding a moon calendar to the top of the vertical tablet, enabling one to deduce the time at night. This dial usually has a compass set into the base or horizontal dial. Another name for this type of dial is "diptych", which is a word given to any two folding leaves with drawings or writings on them.

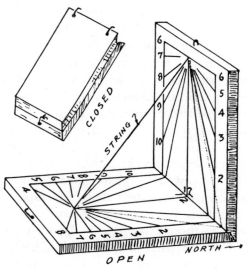

Fig. 10.
Tablet dial

The western hemisphere has not revealed much about its ancient dials. The Mayans and Incas were sun worshipers and we know they constructed sundials, but what varieties and how many kinds were used is not known. Archaeological exploration will no doubt in time uncover their instruments as heretofore in the Mediterranean countries.

Sundials in the United States are really of more recent

origin. When our Pilgrim ancestors landed on these shores they sought a new life in a new world. It is only natural that in a new land new devices such as the clock would be used, leaving sundials behind. In general this is true. Sundials used in the United States during the 18th century are rare,

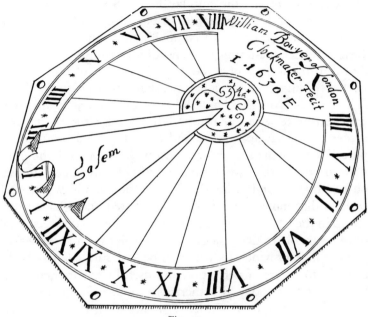

Fig. 11.
First (?) dial in America.

although they were as common in Europe as watches are today.

What is considered the first sundial in America, Figure 11, is now preserved in the Essex Institute, Salem, Massachusetts. This small octagonal dial, about 5 inches across, was made in England in 1630, for John Endicott, who lived in Salem at that time.

Although sundials were uncommon in colonial days, one form appears to have been plentiful—the noon mark. Over the door or on the sill, or by a convenient window a mark was made to record the midday hour. Some of these may still be seen on the older houses in New England. The noon mark has been used in all countries from ancient times in various forms, including the reflection of the sun's rays to a mark on the ceiling of a room; and the spot of light transmitted through a hole in a painted window pane, to the floor.

Today dials in the United States unfortunately are not commensurate with our needs. They are for the most part incorrectly or cheaply made and cannot serve a useful purpose. The few dials in this country which can serve the purpose for which they were intended might almost be counted on the fingers of the hands. This is deplorable when one realizes that most of the dials in the United States have been set up in the last 50 years, during which time no one seems to have been able to devise a dial that has not been constructed before; and previous to which time the sundial had been brought to a state of perfection and usefulness, far exceeding those we set up today. Furthermore those earlier dials were made to tell Standard Time. It is highly improbable that we can devise new types today. Many have tried, only to find its counterpart in some earlier work on the subject, but this does not mean that new designs or applications of the basic principles cannot be devised. We have seen many ingenious designs and applications wrought in the past few years.

The final chapter has not been written, for the past five

years have shown a rapidly growing interest in this oldest of timekeepers. The present generation is eager for hobbies—especially new ones. The sundial will again find its place amongst that group, who will bring back the 20th century dial, both in stationary and portable forms.

II

WHY THE SUNDIAL TELLS TIME

A SUNDIAL will serve you faithfully if it is used. Many dials are not used because the basic principle of their construction and the reason why they tell time are not understood. You can not use your dial to the fullest extent if you know nothing about it other than the fact that it is supposed to tell time—but why?

There is no deep mystery connected with the hour lines on your dial plate or the motion of the shadow cast on it. You will recall from the previous chapter, that at the time of Berosus, the picture of our universe was that of a hollow sphere, with the earth at its center. About 150 years later, 225 B.C., Eratosthenes, an astronomer and mathematician of Alexandria, devised an instrument which consisted of many rings put together in the form of a hollow sphere, with a globe in the center, as in illustration (B) facing page 34. It was made to show the heavens encircling the earth, and it resembled one of those gyroscopes we have all wondered at as children and delighted in playing with as adults. This instrument was called an armillary, which is shown in its simplest form in illustration (A) facing page 34, where only three rings have been used—one to represent the horizon, another the celestial equator, and the third a true north and south line or meridian circle, with an axis pointing to the ce-lestial pole. Sometimes they were made with so many circles they seemed almost unintelligible. However, this jumble of rings is not entirely unfamiliar, for from childhood we have seen them on globe atlases of the world; they correspond to those imaginary lines or circles which divide the earth into the tropic, temperate, and frigid zones. These circles, beginning at the north or uppermost part of the sphere are called—arctic circle, tropic of Cancer, equator, tropic of Capricorn, and the antarctic circle. In addition a sloping ring which cuts across the equator and extends north as far as the tropic of Cancer and south as far as the tropic of Capricorn, shows the path in which the sun and planets appear to move, commonly called the ecliptic.

Such was the armillary. It was used by ancient astronomers for observational purposes; but as time went on it was found to be useful in solving many problems of the sphere. No mention is made of it as a sundial, in early Latin, Greek, and Arabian manuscripts. Vitruvius does not mention it as such, nor does Albategni. This is peculiar for it is the simplest form of a sundial—just a circle. Even as late as the 17th and 18th century it was referred to and used as an instrument to "solve problems of the sphere and to lay out sundials". When it was first used to tell time is not known. The principles of the sundial are more easily understood if the form of the armillary is fixed in the mind in lieu of a working model.

The earthly circles have their counterparts in the sky where they bear the same names, for the heavenly circles are just the prolongation of the planes of the earthly ones. This is easily seen if we mark and name the various circles on the outside of a small rubber balloon representing the earth; then if the balloon could be blown up so large that we could step

in to the center, the circles would be seen from the inside of
the balloon as they would appear in the sky, if the balloon
expanded the same amount in all directions.

You will no longer have to wonder what the rings mean
when you next see this type of instrument, which is often
used today as a sundial. Its application will be obvious as
we proceed.

Our present day method of timekeeping, used in everyday
life, is wholly man made. We have seen fit to make each day
of equal length and divide the day into twenty-four equal
parts, beginning at midnight, with hours running consecutively either from one to twelve or from one to twenty-four.
In the United States we use the former division of the day,
by counting twelve hours from midnight to midday, and
twelve hours from midday to midnight; whereas in some
countries the hours are counted from midnight and continue
consecutively throughout the day to midnight. It is this
division of the day into equal hours that enables us to tell
time by the sundial.

When speaking of circles or parts of a circle, we refer to
them in degrees, a complete circle containing 360 degrees
(360°), each degree containing 60 minutes (60'), and each
minute containing 60 seconds (60''). The degree constitutes
the unit of circular measure. If we divide the celestial equator, which is nothing but a complete circle, into 24 equal
parts, each part will contain 15 degrees. Therefore 1 hour
which is 1/24 part of a circle, will be equal to 15 degrees.

In the construction of a sundial we assume that the sun is
situated on the celestial equator, which is the enlargement
of our earthly equator until it touches the sky. It is also assumed that the sun stays in that position throughout the
year, marking out equal spaces throughout the day and night

equivalent to our twenty-four hours, which is explained in more detail in Chapter Six.

From our geography we know that the equator lies at right angles to the axis of the earth, or the polar axis as it is called. Since the earth's polar axis if extended in either direction would pass through the celestial poles, it follows that the celestial equator lies at right angles to the polar axis of the celestial globe or sphere as shown in the illustration facing page 34. We learn one more thing from observation and geography, that the north star, so-called because it is the star nearest the north celestial pole, is not right over head, nor on the horizon, unless we are situated at the north pole of the earth, or on the equator. We further know that this angle of elevation of the celestial pole above the horizon will be equal to the latitude of whatever place we may be in when we observe it. This is also true for the south celestial pole.

The shadow casting edge of any sundial must always be aligned exactly north-south and make an angle with a level line equal to the latitude of the place for which the sundial is to be used. These two construction principles, together with the assumption that the sun is on the equator throughout the year, are the facts upon which the construction of all sundials is based.

This leads directly to one more fact—that the twelve o'clock line always lies in a true north-south direction whether on a vertical, horizontal, or any other type of dial, because the sun would always be on the north line of any locality at twelve o'clock noon, by sun time. Therefore with these facts at hand it is possible to plot the position and direction of a shadow, cast on any surface in any position. Thus we produce a surface marked out in such a manner that it will record equal hours which may be observed and read

by the position of a shadow. A surface so marked is called
a sundial, and would serve our purpose admirably if we did
not have clocks and watches.

Because the sun does not actually stay on the celestial equator throughout the year, but rather moves along the ecliptic, appearing north of the equator in summer and south
of it in winter, for the northern hemisphere, there is an apparent slowing up or increase in the speed of the sun, which
can be determined in relation to an instrument which records
a uniform speed, such as the watch. This difference in speed
can be observed by comparing the sundial and watch. The
sundial may record noon when the watch says ten minutes
of that hour, or ten minutes after that hour. This is what
your household friend, the almanac, calls 'sun slow' or 'sun
fast'.

The terms 'sun slow' and 'sun fast' were familiar to our
fathers and grandfathers. The sun is said to be slow because
it records the hours after the clock has recorded them, and
is said to be fast because it records the hours before the clock
does. Thus the sun will be slow at some times of the year and
fast at others. The exact difference for any day can be determined, which when subtracted from the dial reading
when the sun is fast and added to it when slow will give
watch time.

We have mentioned the apparent motion of the sun. Do
you remember when you first learned that it was the earth
turning upon its axis that made the sun move? Even a child
in grammar school knows that the earth makes one complete
turn about its axis from the west toward the east, every day,
thus causing the sun to appear to move in a reverse direction
from the east toward the west. It is the earth turning upon
its axis that gives us our darkness and light. Hence, it is ob-

vious that a sundial in Boston would not show the same time as one in New York if both dials were read simultaneously. This is shown in Figure 12, where the real motion of the earth and the apparent motion of the sun are indicated. The two dials mentioned would not show the same time, owing

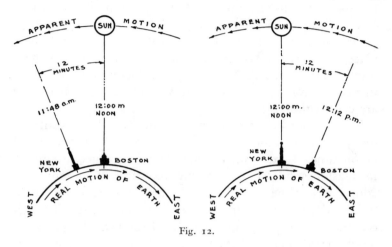

Fig. 12.

to the fact that when the sun is directly over Boston the sundial will read twelve o'clock. It will therefore take some time for the earth to turn sufficiently upon its axis so that the sun is directly over New York, at which time the New York dial will read twelve o'clock and the Boston one will record some few minutes after twelve.

Perhaps a few who read this will recall the days when watches acted the same way. That is, each locality had its own time—called local mean time. Consequently if a New Yorker set his watch at twelve noon and traveled to Boston, he would find that his watch did not agree with that of a Bostonian by some twelve minutes. The Bostonian's watch would be

twelve minutes faster, because he set his, say, when it was
noon in Boston. Since Boston is east of New York, the sun
will arrive overhead in Boston before it does in New York.
The difference in time between the two watches is equal to
the time it takes the earth to turn from that position where
the sun is overhead in Boston to a similar position in New
York as shown in Figure 12.

Or stated in another way—when it is noon at any given
place it is noon at all other places on the same meridian (having the same longitude); and in places having different meridians it is forenoon if they are west and afternoon if they
are east of the given place.

It is not hard to imagine what confusion there would be
today if watches and clocks did not agree. We never would
be on time. In 1884 an international conference assembled
in Washington, D.C., to fix and recommend for universal
adoption a prime meridian, to be used in reckoning longitude and regulating time throughout the world. The acceptance of the prime meridian of Greenwich, England, al-
tered timekeeping. Certain standard time meridians were
designated for everyday use, which varied by one hour, or
fifteen degrees of longitude. In so far as practical, all places
within seven and one half degrees on either side of a standard
time meridian were considered within a certain standard
time zone. There are four standard time zones in the United
States—Eastern, Central, Mountain, and Pacific.

The adoption of standard time made it possible for people
to travel great distances without being at odds in their time.
Only when he entered a different time zone, such as going
from New York to Chicago, would a man have to change
his watch —then he would set it backward or forward ex-

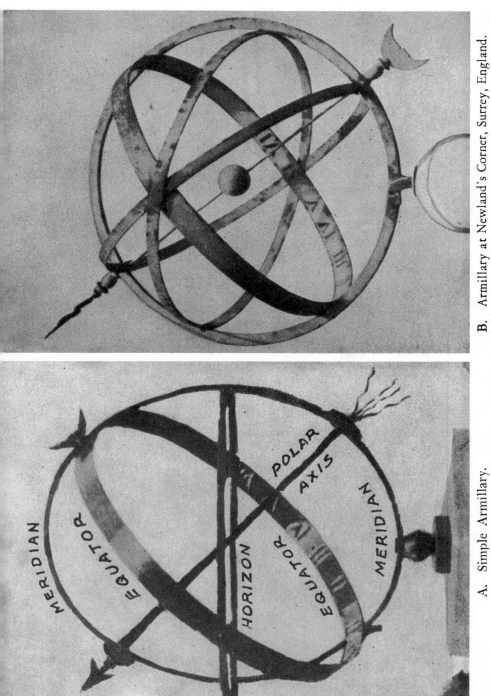

B. Armillary at Newland's Corner, Surrey, England.

A. Simple Armillary.

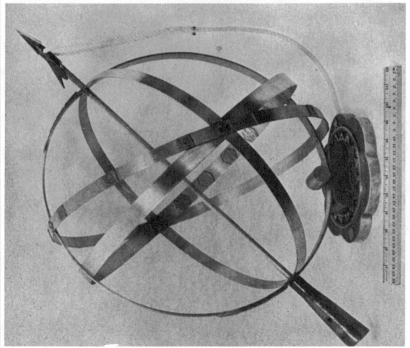

Armillary by Lester M. Peterson.

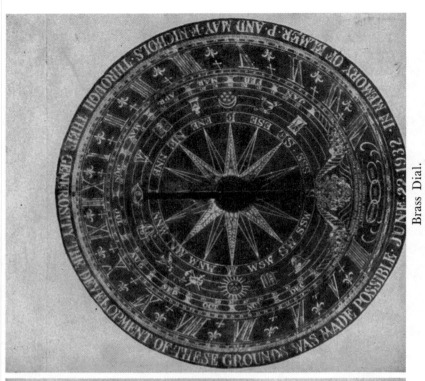

IN MEMORY OF WALTER P AND MAY E NICHOLS THROUGH THEIR GENEROSITY THE DEVELOPMENT OF THESE GROUNDS WAS MADE POSSIBLE JUNE 22 1932

Brass Dial.

actly one hour, depending upon whether he was traveling westward or eastward. Figure 13 shows the time in various cities located in different time zones in the United States, when it is 12 o'clock noon in Boston and New York.

Standard time injected another difference between the sundial and clock which is constant for any particular place, and is derived from the position of any given place in respect to its standard time meridian. For example—Boston is east of

PACIFIC   MOUNTAIN   CENTRAL   EASTERN

SEATTLE   DENVER   CHICAGO   BOSTON
SAN FRANCISCO   EL PASO   DALLAS   NEW YORK

Fig. 13.
Simultaneous standard time in the United States.

its standard time meridian. The sun will be on the standard time meridian sixteen minutes after it has crossed the meridian of Boston. Since neither the standard time meridian nor the meridian of Boston will change, the distance in time between them remains constant. This constant difference is equal to four minutes of time for each degree of longitude. It can be added to or subtracted from the reading of a mean time dial, as the dial is situated west or east of a standard meridian, in order to obtain standard time.

The foregoing shows why a sundial will tell time. The reasons why it can tell time accurately and in accordance with our present day standards are these:

1—We can construct an instrument which will record
equal hours by means of the sun.

2—This instrument can be corrected to show local mean
time in any locality.

3—The local mean time sundial can be corrected to show
standard time.

The subject of time is of such importance to everyone that
a separate chapter has been devoted to it, where tables and
complete instructions for correcting your dial are given.

III

HOW TO DESIGN AND MAKE A DIAL

THERE are several elements that enter the design of a sun-dial beside those which are supposed to tell us just how to go about designing a thing right or wrong, such as symmetry, dynamics, revolution, color arrangement, axes and what not. That's enough to stop any one right there. The design of a sundial is more simple than that. Certain elements are fixed in form and relationship, their differences depending upon the type of dial you are going to make.

As a designer, you must take those basic elements and mold forms and figures that are pleasing to the eye. In doing this you may accentuate the basic predesigned elements or you may hide them entirely.

The best example of accentuation of the basic element is the armillary, where the hours are marked off in equal spaces on the inner surface of a semicircle, or a full circle. The use of other circles, representing the tropics etc., not only accentuates the basic element of the design, but also fixes the appearance and injects a symbolism into the design. As many circles may be used as are necessary to produce the desired result or effect.

A successful designer is much like a successful novelist. We recall an author who has written several books that are

37

classed in what is commonly accepted as good literature; but he was forced to abandon the writing he enjoyed because such books would not sell. He must ever bear in mind those who do not care particularly to read good literature all the time. These people are far more abundant than the purists. Good design is somewhat in the same state. What is considered good by the many may be considered totally wrong by the few who stick to purism. Both sides may be right. What is good design is greatly a matter of personal opinion.

A well designed sundial has certain qualities of a practical value which can be tabulated in the manner of psychological tests—if the answer is YES you credit yourself with a certain number of points,—if your total score is so much, you are a superman,—if less than a certain amount you have no attributes. In designing a sundial or when looking at one already built check off in the list below all of those things which are incorporated in the dial observed, or in your design.

1—Is the dial accurately constructed.
2—Is it properly set up.
3—Is it well made.
4—Can it be easily read.
5—Does it fit well into its surroundings.
6—Is the material of which it is constructed harmonious with other structures in its vicinity.
7—Does it show originality.
8—Is it interesting.
9—Does it tell standard time.
10—Does it impart a feeling of symbolic fitness.

We hestitate to give a numerical rating to these questions, but they have been arranged in order of their importance, based on a census of correspondence, and remarks by many people, who have very definite ideas as to what a good dial should be. That is sidestepping design a bit, but a good sundial that is well designed (in the eyes of many people) must have at least the first six characteristics listed. If any one or all of the remaining four characteristics are also present, such a dial would certainly get a four star rating. You will note that the list is somewhat contrary to the general tenor of this book, because we feel that characteristic number nine is of importance and should be placed among the first six.

The intent of the test is not to lay down hard and fast rules by which a well designed sundial can be produced, but rather as an aid toward that end and that you may better judge your own work and that of others from an impartial viewpoint.

It goes without saying that if a dial is not accurately constructed, and properly set, it just is not a sundial, but an ornament, because a sundial is supposed to tell time.

A dial not well made can never be a source of joy and pride. This refers to craftsmanship, finish, etc., which are essential factors when buying a dial.

Even though your dial is accurately constructed, properly set up and well made, what good is it if it cannot be easily read. The hour lines should be clean cut and open, without nearby distracting ornaments or figures. The illustration facing page 35 shows a dial with so many lines on the surface that it is hard to determine the edge of the shadow or to find the hour circle. In all other respects it is an excellent dial.

No one wants a dial that looks out of place, like a steamship on a desert. A careful selection of materials and location, combined with good taste, will lead in the right direction.

Material may very easily destroy the fitness of a dial. A brick sundial on the side of a lovely old colonial frame house would not look well, nor would a dial with numbers and lines outlined in wrought iron look well just stuck on the face of a brick wall.

Originality expressed in anything is an attribute. Get away from the staid forms of sundials like the horizontal. Make new kinds. Use beams or rays of light. Use the top of an upright pin instead of the conventional triangular gnomon. Add lines which point out the day of the year. Make a dial that does more than just serve you for time. There are hundreds of things you can do to show originality.

If originality is present, interest is almost always bound to follow. Originality and interest may sometimes overshadow a lack of fitness. Such a dial is that at Queens' College, Cambridge, England, see illustration facing page 42, which is painted on a masonry wall. The main points of interest are the signs of the Zodiac in color, and the table below the dial which enables one to tell time by the light of the moon. This is the famous moon dial mentioned in many reference books.

A dial should tell standard time if it is to be of any practical use today. Therefore this becomes a requisite, except in special instances, where symbolism may be the important factor, or in memorial dials. But symbolism need not be destroyed by constructing the dial to tell standard time.

Symbolic fitness is not confined to memorial dials, or those of monumental character. Symbolism may be present in your own sundial, by injecting into the outline or ornamentation the figures or characters symbolic of your hobby or profession, as is often done in bookplates.

An interesting example of symbolic fitness is shown in Figure 14. This dial was designed by a man of literary habit, who owns a stone cabin in (yes, in) Seneca Lake, New York. Here is what he says about it—"This dial is to be made of

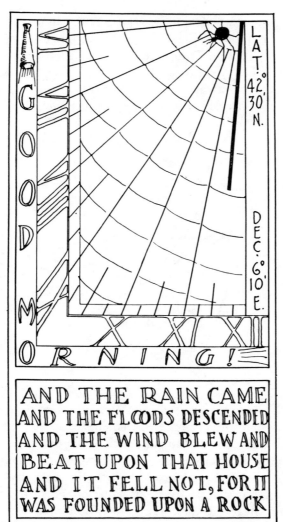

Fig. 14.
Symbolic fitness.

sandstone and set into the wall of a rock cabin which I am building as a sort of 'island' just off the shore cliffs of Seneca Lake with its foundations resting solidly on a submerged stratum of Devonian rock. Because of the high cliffs immediately to the westward, the sun sets here at noon and the dial is therefore provided with morning hours alone." The hearty

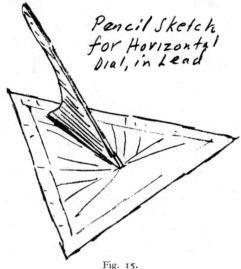

Pencil Sketch
for Horizontal
Dial, in Lead

Fig. 15.

welcome "Good Morning" doubtless alludes to cheeriness within despite the presence of the sinister spider, while the symbolic quotation in the panel below the dial refers to the situation of the cabin among the waves and on a solid rock foundation. Another stone, on which has been carved the corrections for standard time, is to be placed in the wall of the cabin, near the dial. We give this sundial a four star rating.

The design of a sundial should include its support. Treat

Wall Dial, Queens' College, England.

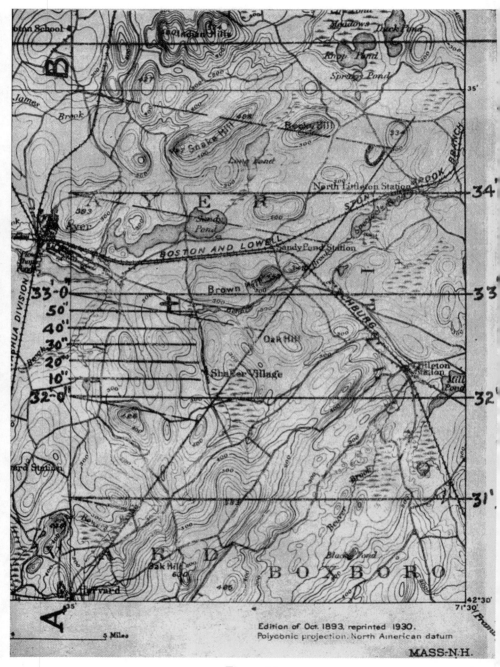

Figure 16.

the dial and its support or pedestal as a structure. Have the various parts—gnomon, dial plate, etc., blend into each other so that nothing about the whole will seem incongruous.

How to Make a Dial

The first step in making a dial is to determine what kind of a dial will best fit the place where it is to be set up, and the kind that will best serve your purpose.

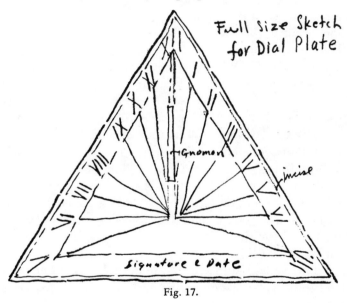

Full Size Sketch for Dial Plate

Gnomon

incise

Signature & Date

Fig. 17.

The second step is to design the dial and its support by sketching it on paper, Figure 15. Incorporate in the sketch only those things which are necessary. Save time. Do not put in too much detail at first. When the general outline in sketch form has been approved, roughly work up your layout at full size, Figure 17, and add more detail. Do not try to lay out the hour lines except in a general way.

After the full size drawing has been brought to the point where you are ready for the hour lines, the next step is to lay out those lines, but before this can be accomplished the latitude of the place where the dial is to be situated must be found, because the shape of the gnomon is dependent upon it. This can be done very simply as follows:—

Obtain a copy of the topographic map for your locality; a portion of one of these is reproduced as Figure 16 facing page 43. These maps are produced by the U.S. Geological Survey. Persons east of the Mississippi River should write to U.S.G.S., 1200 S. Eads St., Arlington, Virginia 22202; west of the Mississippi to U.S.G.S., Denver Federal Center, Bldg. 41, Denver, Colorado 80225. The most useful maps cover an area of either 7½ or 15 minutes of latitude or longitude on a side. The latitude may be obtained from the horizontal lines and the longitude from the vertical lines. The bottom and top lines are marked at either end with the degree and minute of latitude, and the intermediate lines are designated by minutes only, except where a change in the degree takes place in the middle of the sheet. The vertical lines of longitude are similarly marked.

Now find your position on the map and make a cross as shown in illustration facing page 43. In the figure, the cross lies between the latitude of 42° 30′ and 42° 35′; therefore we must find how far the cross is away from either parallel of latitude in order to find the latitude of the cross. This can be accomplished by using the line of longitude, AB, or by using a line which passes through the cross and perpendicular to the lines of latitude. In either case another line is drawn at random from A to C, as shown. Lay off five equal spaces on

WESTPORT, CONNECTICUT

This dial was made for a yachtsman's garage, which stood about 20 feet from the sea, so that if his grandchildren were out sailing, they could tell when to come in for lunch. The gnomon is in the shape of a sextant, and a correction table (not shown) enables standard time to be determined.

this line beginning at A, ending at C. Then join B and C with a straight line. Draw other lines parallel to BC through the points on AC and cutting AB. This is shown by the dash lines in the figure. Thus will the distance between the two given lines of latitude be divided into five equal spaces, each equal to one minute. The latitude may be easily obtained

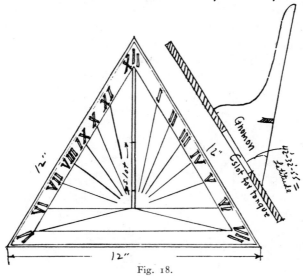

Fig. 18.

within a few seconds of arc by further subdividing the minute space into six equal parts of ten seconds each.

The cross lies about half way between the line of 32′50″ and the line of 33′00″. Thus, the latitude at the cross is 42°-32′55″. The longitude is 71°33′45″, which is found in a similar manner by using the vertical lines, subdividing the five minute interval into a sufficient number of spaces to obtain one minute intervals, and so on.

The latitude must be found in order to lay out the hour

lines. The longitude must be found if the dial is to be used for standard time; to find the meridian; or to find the declination of a plane.

When the hour lines have been laid out, place them roughly in the position you want them, on the full size sketch; then make a careful plan of the dial plate and gnomon at full size, either on fairly heavy tracing (transparent) paper, or on a good sheet of white paper. This drawing is used to transfer the design to the dial plate; or if you prefer, lay out the final plan on the dial plate. It is better, however, to retain a careful paper copy of the hour lines and design, for reference.

Those who have a mathematical bent will probably want to compute the hour lines. Complete instructions for doing this will be found in Appendix I. Unless you are adept at handling logarithms you will probably refer to Chapter VII for the construction of the hour lines for your dial. Computation is by far the best way to lay out any dial, but the same accuracy can be obtained with a large, well made protractor.

The protractor is an instrument used for laying off angles in degrees and fractions thereof. If you do not already have one, an accurate machine ruled paper protractor can be purchased from drawing and engineering supply stores for a nominal sum, twenty to forty cents. However, you may prefer a plastic protractor, which is more durable and also comes in several sizes with different graduations. Large protractors will generally have the finest graduations. One that has fifteen minute (quarter degree) marks will enable you to plot angles as accurately as needed to make sundials, and is otherwise useful to have around. Thirty minute intervals are the largest divisions you should use. Both of these types are shown in Figure 19. The protractors may be purchased as 180° semi-circles or as 360° full circles.

Scale Model for Horizontal
Dial.

Completed Lead Dial.

Brass Horizontal Dial by I. Klocksperger, Bohemia, 18th Century.

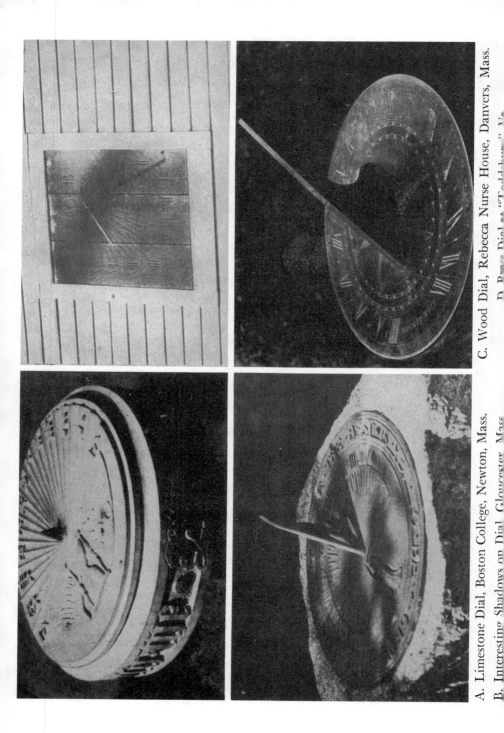

A. Limestone Dial, Boston College, Newton, Mass.

B. Interesting Shadows on Dial, Gloucester, Mass.

C. Wood Dial, Rebecca Nurse House, Danvers, Mass.

D. Brass Dial at "Toddsbury," Va.

When using a protractor, always draw lines long enough to extend beyond the limit of the protractor. A large protractor will minimize errors. Figure 20 demonstrates the successive steps in laying out an angle with the cardboard instrument; in this case, an angle of $42°30'$.

With our latitude known and the hours laid out, we can continue with the construction of the dial shown in Figures 15, 17 and 18. A medium soft piece of lead twelve inches square and one quarter of an inch thick was obtained from a hardware store at a very nominal cost. The careful drawing of the design was placed on the lead plate and held firmly in

GRADUATED IN 30' INTERVALS

8" CIRCLE

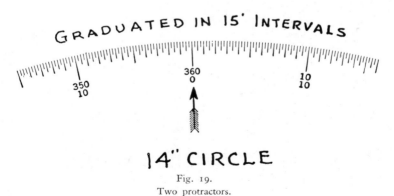

GRADUATED IN 15' INTERVALS

14" CIRCLE

Fig. 19.
Two protractors.

place while the extremities of the lines were pricked into the metal with the sharp point of a round ice pick found in the kitchen table drawer. The lines could have been transferred by placing a piece of carbon paper beneath the drawing, and going over them with a sharp hard pencil.

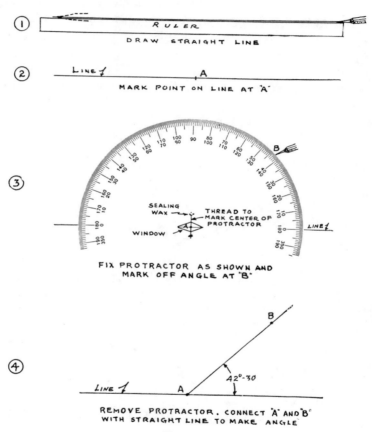

Fig. 20.
Steps in laying out angle with protractor.

After the design and lines have been transferred to the dial plate, they must be made permanent, prior to which the plate was cut in the triangular shape outlined in the sketches, by using a small, thin, fine tooth saw with a narrow blade such as may be bought in the ten cent stores. Cutting the lines in the lead was accomplished by first scratching them on the surface with the ice pick guided by a straight edge ruler. Do not bear on too hard when scoring the lines for the first time, a light pressure is sufficient.

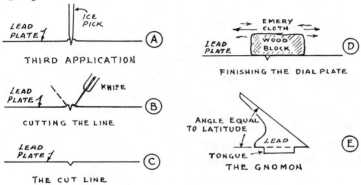

Fig. 21.
Cutting the lines and finishing.

The rough edges of the triangular plate were smoothed off with emery cloth and a cheap file, after which the ice pick was applied to the lines a second time and then a third; each time more pressure was exerted. The third application of the pick gave sufficient depth and had raised the lead on the side of the groove as shown at A, Figure 21. A pocket jackknife was then brought in to play to cut off the raised portion, on both sides of the lines, and at the same time giving the desired V-shape to the cut by holding the knife at a slight angle, as at B. The completed line is shown at C. After all the lines were cut in this manner a piece of emery cloth was wrapped around a square edged block of wood and rubbed lightly over the surface to take off any raised portions or roughness at the edge of lines as at D. The numerals were cut in the same way, by first scoring with the ice pick and then shaping with a knife, and finishing with emery cloth.

With the hour lines and numerals thus cut in the plate, a slot was cut out just wide enough to allow the passage of the tongue at the bottom of the gnomon. This completed the work on the dial plate. The gnomon was cut out and shaped by the same tools used on the dial plate. The completed gnomon is shown in Figure 21. It was cut out of one of the waste pieces of lead removed in shaping the dial. The curve was cut out with the saw used on the plate. The gnomon was carefully set in place and soldered in a perpendicular position. Two blocks of wood, 2″ x 4″, were held on either side of the gnomon while soldering. The finished dial is shown in the illustration facing page 46. When set up it was found to be entirely satisfactory and accurate. A table was made to show the correction to be applied to obtain standard time.

One need not delay making a dial until he can obtain expensive tools. A good dial can be made with inexpensive equipment, or with the tools at hand in most every household. Each individual has his own method of doing things— what may be easy for one, may be hard for another. Therefore we have only shown what can be done with simple tools, in the short space of an afternoon.

There are so many methods of attaching the gnomon to the dial plate and the dial to its support, all of which are satisfactory, that it would be folly to suggest any one or two. In any case the gnomon should be firmly attached in its proper position so that there will be no "play" if the apex is touched.

IV

SELECTING THE DIAL TO MAKE OR BUY:

GUIDELINES AND MATERIALS

As MUCH care should be used in selecting the kind of dial to make or buy and its material, as would be used in selecting your new automobile or dress. More so if you are buying a dial. When milady has a dress made she is particular to make known just what she wants—it must be the right type for the purpose; it must fit her, not someone else; and it must be well made. If any one of these things is lacking the dress is not satisfactory nor acceptable.

The same is true for a sundial—it should be the right type, that is, vertical for a wall, an armillary for educational purposes, or if the time of sunrise and sunset is desired it must have the necessary lines upon it. A dial should be constructed for the place in which it is to be used, not some other place; and it should be well made. If the sundial lacks any one of these things it will not be satisfactory, nor acceptable.

The first thing to do in selecting a dial is to determine the class or type of dial you think would look well in the spot picked out for it. All dials can be placed in one of the following four major classes:

I—Spherical
II—Conical

III—Plane

IV—Portable

These major classes may be further subdivided into various kinds or forms, which generally are designated by names descriptive of their location, form, or the position of the surface upon which the hour lines are drawn. For example, an equatorial dial is a plane dial whose surface lies parallel to the plane of the equator; a horizontal dial is a plane dial whose surface is level; and a spherical dial may be a segment of a sphere or a complete sphere.

Class I consists of the hemispherium, the sphere, etc. They may be hollow, solid, concave, or convex.

Class II comprises those dials that are segments of a cone or a complete cone. They are rare and of ancient origin. There is no good reason why they should not be made today. The so-called "goblet" dials that look like a wine glass with a vertical pin in the center, are conical dials.

Class III consists of two major subdivisions—attached and detached, which include all stationary dials drawn on plane or flat surfaces. Attached dials are those affixed to or made a part of some structure. Detached dials are free standing.

Class IV might very well be considered a major subdivision of each of the foregoing classes. It is separated because only those dials are included whose function and relation to others is comparable to that of the watch to the clock.

There are innumerable kinds of dials, which generally fall in either Class I or III, but each class offers a great variety in form and position.

After the class, type, or kind of dial has been decided upon, the next step is to make sure the dial is figured for your latitude. The only real way to determine this, if there is any doubt, is to check it yourself. It is not sufficient to check only the slant of the gnomon. If you are buying a dial the position of the hour lines should be checked by making a small drawing of them and the angle of the gnomon, in accordance with the method outlined for that dial, in Chapter VII. Compare your drawing with the dial in the store.

Most store salesmen will know less about dials than you do. If you buy a dial which has not been designed and computed for your place, it can be adjusted by studying the construction of reclining dials in Chapter VII. Therefore if a horizontal dial is not laid out for the proper latitude it is not a serious thing for one who is willing to determine what adjustment should be made. Otherwise, don't buy it.

The next step and perhaps the most important one of all is that the dial be well made, not only as to craftsmanship, but as to the accuracy of its construction. This will be shown when the check mentioned above is applied.

Right here we wish to caution you against buying cast dials. These dials are made from a mold by the hundreds, are shipped around to various stores, and generally sell at low prices. They can usually be spotted because the shadow casting edge of the gnomon is often round, when it should have a sharp, clean cut edge; the hour lines are frequently excessively wide; the dial plate may be scooped out when it should be flat; the hour lines are sometimes raised and the gnomon may not even be properly placed. We know of one instance where the angle of the gnomon was changed for various locations, but the hour lines remained the same; therefore you cannot rely on checking the gnomon alone.

Do not be fooled. You cannot buy a good dial at a cheap price, except in very rare instances. For that matter you can have some types of dials made to order as cheaply as you can

buy a well made stock dial. A good stock dial is often carried
for a long time, thus the price is usually high.

There are certain features that should be apparent in any
well made dial. The dial surface upon which the hour lines
are drawn should be smooth and flat, the lines clean cut and
straight. Excessively wide lines are not indicative of accuracy. Raised lines and depressed surfaces are usually incorporated in dials which were never intended to tell time.
Make sure the shadow casting edge of the gnomon is sharp
and straight. Take the dial into the sunlight and see if the
shadow cast is well defined, and straight. Touch the apex of
the gnomon to find out if it is firmly fixed in place. A loose
gnomon is no good. Can the position of the shadow be easily
read on the hour circle?

Don't forget that no matter how good your dial may be it
will be useless if not set up properly. Full details for setting
the various dials are given in Chapter VII.

We repeat that the selection of a dial is dependent largely
upon the owner's needs and the place in which it is to be located. There is little reason for placing a horizontal dial in a
sheltered place that has sunlight only a few hours during the
day, when upon the side of the house there is a large blank
wall that receives full sunlight all day and can be easily seen
from the garden—an ideal situation for a vertical dial. Then
again, if one wishes a dial to tell the time of sunrise and sunset throughout the year, the pillar dial or some other combination will best serve the purpose. If there are children in
the family, an armillary would be best from the educational standpoint, for hours can be spent demonstrating
various celestial and terrestrial facts. The armillary is an
excellent instrument to help father over the rough spots
caused by Junior's questions.

If some means of checking your watch is desired other than by calling the telephone company, you would be certain to enjoy most of all that extremely accurate instrument — the heliochronometer.

The following notes on the various dials are intended only as a guide, to help determine the kind of dial best suited to your particular needs. For convenience each dial is treated separately and reference is made to the corresponding construction diagram in Chapter VII.

The Equatorial Dial (Plate I)

The equatorial dial derives its name from the position of the dial plate, which lies parallel to the plane of the earth's

equator. It has a perpendicular gnomon pointing to the celestial pole. It may be moved from place to place without refiguring, because the construction of the hour lines is not dependent upon the latitude of the place where it is used.

Because the sun moves in a path north of the equator during the summer months and south of the equator during the winter months, the dial should be in the form of a thin plate, with hour lines inscribed on both the upper and under surfaces in order for it to show the time throughout the year.

It is an interesting dial, easily constructed and not common. It makes a splendid accent or focal point in the garden or on the lawn, where there is little interference from nearby buildings and trees. It can easily be adapted to standard time and used as a heliochronometer.

The Horizontal Dial (Plate II)

This is the common variety sold in the stores and found in 99 and 44/100% of the gardens. It lends itself easily to freedom in pedestal design and decoration of the dial plate, which can be as elaborate as the owner chooses.

In order to be at all satisfactory as a timekeeper it must be figured for the locality in which it is to be placed. When

situated in an open spot in the garden or on the lawn it will tell time from sunrise to sunset throughout the year.

Like the equatorial dial, it may be used as a focal point.

Although it is usually wise to place a sundial where it will receive full sunlight all day, the lower left photograph (B) facing page 47, shows the attractive effect gained by placing the dial at the south edge of a border of trees, where the sunlight filters through the outer branches, and the edge of the shadow of the gnomon is glimpsed by the constant play of light and shadow across the dial plate.

The Direct Vertical Dials (Plates III, IV, V)
and the Pillar Dial

There are four direct vertical dials, one for each of the cardinal points of the compass—north, east, south and west. They lie perpendicular to the plane of the horizon and they must face directly toward the cardinal points. They must also be made for the place in which they are to be set up. Each dial is limited as to the number of hours the sun will shine upon it. Except for the few months of summer between the equinoxes (when the sun is north of the equator)

the sun will not shine upon the north dial between 6 a.m. and 6 p.m. The east dial will show the time from sunrise to noon; the west, noon to sunset; the south from 6 a.m. to 6 p.m.

Few walls or other upright objects face the true north, south, east, and west points of the horizon; therefore these dials are usually combined on one block forming what is known as the pillar dial.

The pillar dial will show the time from sunrise to sunset throughout the year. It can also be made to show the time of sunrise and sunset. This dial deserves a prominent position and ideally meets the requirements of a large formal garden. Such dials were often placed in the town square, where the passerby could obtain more information from its faces than we can obtain from its successor, the town clock. Many pillar dials may still be found in the rural districts of England, Scotland and Wales.

THE POLAR DIAL (PLATE VI)

The surface of the dial lies parallel to the plane of the earth's polar axis. Like the equatorial dial, the construction of the hour lines does not depend upon the latitude of its situation. It shows the time from 6 in the morning to 6 in the evening. This dial is not common and is seldom used except in combination with other dials.

THE DECLINING DIAL (PLATE VII)

There are four kinds of declining dials —those facing the south, declining toward the east or west; and those facing the north, declining toward the east or west. They do not face the cardinal points of the compass and are exceedingly well adapted to use on walls of houses, churches or other buildings. The length of time the sun shines upon them varies according to the amount of their declination. They may be plain and simple or monumental in character.

DIRECT RECLINING DIALS (PLATE VIII)

There are four direct reclining dials; each faces a cardinal point of the compass; and their planes (as you stand before them) lean from you (recline from the zenith). The equatorial and polar dials are reclining but due to their position they are named separately. The length of time the sun shines

upon them varies according to the amount of their reclination or slope. They are seldom used except in combination with other dials, but there is no reason why one could not be placed on the roof of a summer house or small shed, if the roof faces one of the cardinal points.

DECLINING-RECLINING DIALS (PLATE IX)

This type of dial was commonly used during the Renaissance period when blocks of stone were cut with a great many faces, like a gem; some with as many as seventy-two surfaces. These were called facet headed dials, but a better reference would be multiface dials. The great number of gnomons protruding from all sides made them look like octopuses. However, these dials can be placed on sloping surfaces that do not face the cardinal points of the compass, such as most roofs, or the sloping cap of a garden wall. They are not so easily

laid out or computed, and seldom does one take the trouble
to make them.

The Armillary (Plate X)

The armillary is one of the most flexible dials, from the
standpoint of design; and it is the best from the educational
point of view. Every first-year student in astronomy becomes
familiar with its principle and form. The several rings por-
tray the major circles of the earth in their proper relation to
each other, the whole being symbolic of our universe. A
small ball placed on the gnomon at the center of the sphere
represents the earth, making this dial a veritable primer of
astronomy. Children take great delight in it and father generally is thankful for it, because it serves so well to answer
many questions put to him by the youngsters.

Unfortunately the armillaries generally look out of place,
because they are poorly designed—lack scale. However, it
has no peer as a memorial, one of the finest of this type is
described and illustrated in Chapter XIV. The use and function of the armillary is so important that it is mentioned frequently in these pages.

The Analemmatic Dial

This dial has been included by popular demand. More
often than not when one finds how much work is entailed
in making it, another type more easily constructed is substi-

tuted. However, the chief use of an analemmatic dial, is in conjunction with a portable horizontal dial to enable it to be set properly without the use of a meridian line.

This is accomplished by placing both dials on the same plate. The horizontal dial uses a sloping gnomon which records the hours in respect to the distance of the sun from the meridian, whereas the analemmatic dial makes use of a mov-

able vertical pin which can be set for each day in the year, thus recording the time with respect to the azimuth of the sun (angular distance on horizon between south point and the foot of the perpendicular from the sun to the horizon).

If the combination dial is placed on a level surface, the pin on the analemmatic dial set opposite the proper day, then the whole turned so that the time shown by both dials agrees, the horizontal dial will be in its proper position.

THE HELIOCHRONOMETER

The most accurate dial made. Time may be obtained to within a few seconds. The commonest forms are the equatorial, half cylinder, and other types where the hours are marked in equal spaces on the surface of the dial. Much originality can be used in their design, particularly in the arrangement of the mechanical parts. This type of dial always attracts a great deal of interest wherever set up. They are essentially standard time dials, and can generally be easily adapted to daylight saving time. Perhaps you can turn your present dial into a heliochronometer by adding a vernier as described in the following chapter. Further instructions are also given in Chapter VI. We urge their construction.

When one builds a house or garden, everything must be just so; but look about you, in your neighbor's garden or on his house—when it came to the sundial, which is often of great importance in the garden scheme, any old thing was put up. Don't blindly select a sundial to make or buy.

Materials

The material out of which to fashion a sundial presents no problem to the craftsman; but if one is not endowed with a set of woodcarving and engraving tools he must select that which is best suited to his skill and the tools at hand. The usual procedure is to get the proper tools for the material. Why not put the cart before the horse and get the material for the tools at hand. Because the tools at hand were poor, medium soft lead was used to make the dial described in Chapter III.

The materials commonly used in making sundials are well seasoned wood, metal, and stone. Lead, brass, and bronze comprise the metal group. Granite, limestone and marble are the more common stones. Before we go into more detail about these materials, consider the factors that more or less govern their selection.

In addition to appearance, the ease with which a shadow can be seen is of vital importance. Background color affects this more than anything else. Certain colors such as those just off the white-cream, light gray, and so forth, will catch the shadow when there is very little sun; whereas the dark gray of lead, the natural weathered finish of brass or copper,

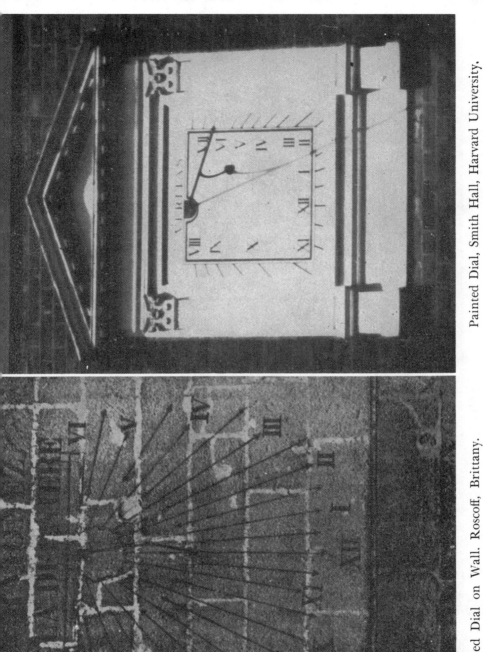

Painted Dial on Wall. Roscoff, Brittany.

Painted Dial, Smith Hall, Harvard University.

Vertical-Declining Dial of the Foxboro Co. in Massachusetts.
24-hour Dial at Thule, Greenland.　　　Roman Hemicycle, about A.D. 300.

or lacquered brass or copper need brilliant sunlight. While this may seem of relatively little importance, in some dials it is all important. To the novice, workability is another factor which affects the choice of material.

Not many dials are made of wood, yet it is a serviceable material that goes well with many others, and is easily "worked". Wood, after a short exposure to the weather, has an appearance of age that is hard to imitate by artificial means. It is admirably suited to an old house. Note the dial on the Rebecca Nurse House in Danvers, Massachusetts, see illustration facing page 47. This house was built about 1658, but the dial was placed over the door only a few years ago. Wood was quite popular in the 16th and 17th centuries. Whitewood, white pine, and hardwoods may be used with success. Such dials may be carved or painted. The gnomon is generally made of metal— copper, brass, bronze, or wrought iron; that on the Nurse house dial is a wood dowel that has weathered the storms without visible damage. The illustration facing page 62 shows a painted wood dial with iron gnomon, on the south wall of Smith Hall, Harvard University.

Next to wood, lead is most easily "worked". It comes in three hardnesses—soft, medium soft and hard. Hard lead should be purchased, because the other two grades will not hold their shape without backing, and they have a tendency to curl at the edges in the spring, if left out all winter. The gnomon can be of the same material, but brass or bronze will be more serviceable. An accidental touch of the hand is apt to bend a lead gnomon.

The use of lead dials should be confined to well protected places, for they are easily damaged. Lead has a very pleasing color if allowed to weather naturally.

Brass is a hard metal, which requires care and special tools to work it. If you choose this metal, do not ask for ordinary brass, but specify "engraver's brass", which contains a little lead making it easier to handle. Engraver's brass is not soft, neither is it too hard. Unless the lines are to be cut in deep, no special tools will be required. The same material may be used for the gnomon. Natural weathering gives a pleasing effect, which can also be obtained by burying the completed dial in a pile of well-rotted manure for a few days. Brass dials are common, durable, and not easily damaged. They may be used in unprotected places, but the gnomon should be firmly attached to the dial plate, which in turn should be firmly anchored to its support. Many beautifully engraved memorial dials have been made of this material. The illustration facing page 47 shows an old brass dial at "Toddsbury" in Virginia.

Bronze is another hard metal admirably suited to dials in unprotected places, but it requires more experience in handling than any of the foregoing materials. Specify "sheet bronze" when ordering. Bronze is better adapted to large dials of monumental character.

Stone is also used more often for memorial and monumental dials, than for personal dials. Making a dial in stone requires a lot of patience and frequently causes one to give up. Granite requires special tools as do other stones of similar character. Limestone, sandstone, and slate are comparatively soft and more easily "worked" by the novice. Limestone is the best of the three, for in addition to workableness, it has the advantage of a good background color to catch the shadow. Slates and sandstones are not so serviceable because of their tendency to chip and flake while cutting or on exposure to the weather. If you insist on making a dial in stone we recommend taking a sample piece to a commercial stone cutter, and ask him for advice as to the best way to cut the lines in that piece. He will gladly help you, show you the tools needed and where to obtain them. Or if you are stubborn there is nothing to do but struggle with the job and if at first you don't succeed, try again. We know of one man who bought a small electric handy tool (quite inexpensive, about $15) with a variety of abrasive wheels and drills, which worked swiftly without chipping on the softer rocks. If ordinary rock cutting chisels, drills, etc., are used with a hammer or mallet, you will save much grief by pasting a carefully drawn plan of the dial on the stone. The drawing should be made on a not too heavy soft paper; then the lines are cut into the stone, through the paper. The paper will retain the outline of the design until it is cut, and acts as a safeguard against chipping.

The limestone dial shown in the illustration facing page 47 was recently given to Boston College in Newton, Massachusetts, by a stonecutter who is a sundial fan. Note the small conical bronze gnomon, and the shallowly scalloped dial plate, to bring out the lines. The position of the shadow cast by the top of the gnomon alone indicates the time. The simple design, ably executed, and the conical gnomon, have been sufficient to create much interest in this otherwise ordinary horizontal dial.

Many interesting dials have been painted on wood and masonry walls. Simplicity is the keynote of the dial, shown in the illustration facing page 62, painted on a wall in Roscoff, Brittany. Contrast this with the dial painted on the wall at Queens' College, facing page 42. Paint is an excellent material for dials. Its use makes the construction of a dial easy. A wall dial plate can be made in the cellar, painted there, and later attached in its proper place. No carving, no worry. If you make a mistake, it is easily rectified. A painted dial will last a lifetime, as evidenced by many old ones to be found in England and European countries. Only one word of caution is necessary for painted dials—buy good paints. They cost a little more but are well worth it in the end. There are many reputable brands on the market, one as good as the other.

If paint is to be used on brick or masonry walls, the area covered by the dial should first be prepared to receive the paint, to avoid flaking off or peeling within a short time. This can be done by first washing the surface with a dilute solution of muriatic acid, a sufficient quantity of which may be procured from your druggist or a chemical supply store. When the washed surface is thoroughly dry, apply a coat of raw linseed oil. The oil enters the pores of the stone or brick and when dry is hard and impervious to weather. One coat of oil may be sufficient, but a second coat will be certain to give a good base to which the paint can be applied.

No elaboration on the method of preparing wood by first using a paste filler, then shellac, then paint, need be made here. It is always well to read the directions on your can of paint before using it. One manufacturer may suggest a different preparatory method for his particular paint.

Painted dials will frequently be found to better suit the situation than one made of another material. The gnomon can be iron, brass, or bronze.

Some thought should be given to the selection of material for your dial. If you are particular about other parts of the dial, be particular with its material.

V

PARTS OF A DIAL YOU SHOULD KNOW

K NOWLEDGE is understanding—to know, understand, fully appreciate, and get the most out of your dial, you should become familiar with the use, location, and names of the various parts commonly associated with sundials. Some terms have not been fully described in the text, in order to preserve clarity and continuity. They are listed and defined here for reference.

CENTER OF DIAL—The point where all the hour lines meet; or where they would meet if extended.

DECLINATION OF THE DIAL—This refers to the angle formed by the intersection of the meridian line (true

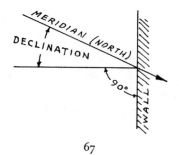

north–south line) and a line perpendicular to the surface of the dial upon which the hour lines appear. The angle is always measured from the south toward the east or west, or from the north toward the east or west. The term is applied to those vertical or inclined (sloping) dials which do not face the true north, south, east or west points of the compass.

DIAL PLATE—The surface or plane upon which the hour lines are laid out.

DIAL WITHOUT CENTER—This term is applied to those dials upon which the hour lines do not meet at a common point. Vertical dials facing the east and west points of the compass are typical, their hour lines being drawn parallel to each other.

FOOT OF THE PERPENDICULAR STYLE—That point on the substyle where the perpendicular style is erected.

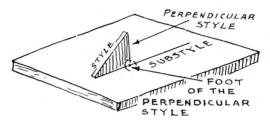

GNOMON—Any object which, by its shadow, serves as

an indicator. One of the most ancient instruments for telling time. An obelisk or staff set perpendicular, the time being recorded either by the angular movement or length of its shadow. The style of a sundial is often erroneously referred to as the gnomon. See STYLE.

HEIGHT OF THE PERPENDICULAR STYLE—The

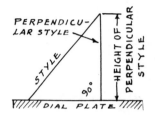

distance measured from the top to the foot of the perpendicular style.

HEIGHT OF THE STYLE—The angular, or linear, dis-

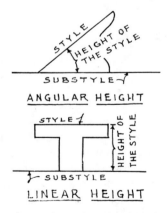

tance of the style above the substyle. The linear distance may be measured in inches, millimeters, or any standard division.

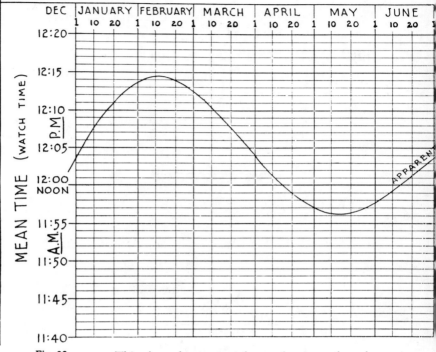

SCALE OF DAYS
1 5 10 15 20 25 30

RELATION BETWEEN APPARENT
COMPUTED FOR THE 75ᵗʰ MERIDIAN

Fig. 22.

This chart shows, at a glance, the time when the sun will be on the standard meridian (at the left), and the equation of time (at the right), for any day in the year. A correction will have to be made for the observer's meridian, if his meridian is east or west of the standard meridian for the time zone in which he is stationed. This correction amounts to four minutes for each degree of longitude east or west of the standard meridian. If the observer's meridian is east of the standard meridian, the correction must be subtracted from the time shown on the chart; if west, the correction must be added. *Example:* Find the time the sun will be on the meridian at Boston, Mass., on March 20. Accord-

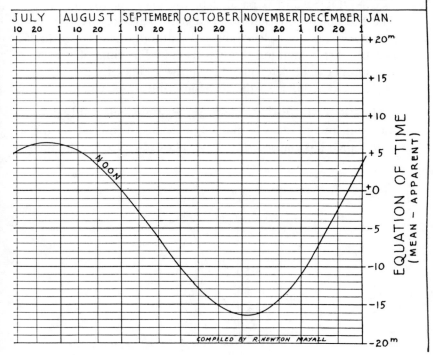

NOON AND MEAN TIME
(EASTERN STANDARD.TIME)

COMPILED BY R. NEWTON MAYALL

ing to the chart the sun will cross the 75th meridian at 12h 7.5m P.M., E.S.T. The longitude of Boston is 71.07 degrees. The difference between Boston and the standard meridian (75th) is 3.93 degrees. Applying the correction of four minutes for each degree of difference, 15.7 minutes must be subtracted from the time obtained from the chart, since Boston is east of the standard meridian. Therefore the sun would be on the meridian at Boston, March 20, at 11h 51.8m A.M. E.S.T. Note that the equation of time is equal to the mean time minus the apparent time. This chart is applicable to the standard meridians (see upper right hand corner). See also the table on page 87.

HORIZONTAL LINE—A line drawn on the dial plate at its intersection with a plane which passes through the nodus (see NODUS) and is parallel to the plane of the horizon. The lines of declination do not extend beyond this line. It is, however, seldom placed on dials, although it plays an important part in making the dial useful. The time of sunrise and sunset can be determined from the points where the lines of declination intersect the horizontal line. See Chapter IX.

HOUR ANGLE—The hour angle of any celestial body—star, sun, moon, etc.—is that angle or arc measured by the time which has elapsed since it was last observed on the me-

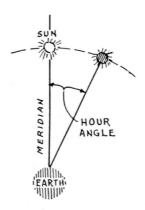

ridian. Since this angle depends upon time, it is usually measured in hours and minutes, instead of degrees. One hour equals 1/24 part of a circumference, which contains 360°; therefore 1 hr. = 15°, 2 hr. = 30°, etc. See conversion table, Appendix II.

HOUR LINES—The lines described on any surface for telling time.

MERIDIAN—A great circle of the celestial sphere passing through its poles and the zenith of a given place. A great circle on the earth passing through the poles, and any given place.

MERIDIAN LINE—On sundials, the 12 o'clock line, which must always lie in the plane of the meridian. On the earth, a true north–south line.

NODUS—A specific point on the style whose shadow traces out the path of the sun. In order to trace or show the path of the sun on any particular day, the shadow of the whole style cannot be used; therefore some point must be selected, which may be at the apex of the style or at any other location along its length that may be convenient. The

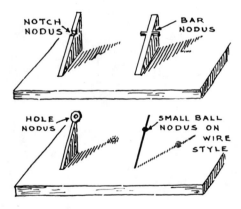

nodus should be designated in such a manner that its shadow can be easily seen on the dial plate, in sharp contrast to the shadow of the whole style.

PERPENDICULAR STYLE—A line drawn through the nodus to the substyle, and perpendicular to the dial plate.

PLANE—A flat, or level, surface.

PLANE OF HORIZON—A plane which if extended would intersect all points of the horizon.

RECLINATION OF THE DIAL—This term is applied to those dials which occupy neither a vertical nor a horizontal position, and it refers particularly to that angle, measured in degrees, formed by the intersection of the dial plate with a line perpendicular to the plane of the horizon.

STYLE—That edge of the gnomon elevated above the dial plate which casts the shadow by which the time is recorded.

SUBSTYLE—The line upon which the gnomon is erected. The base of the gnomon.

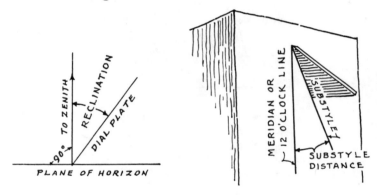

SUBSTYLE DISTANCE—The angle which the substyle makes with the meridian or 12 o'clock line. See construction of declining dials, Chapter VII.

VERNIER—A short scale named after its inventor, Pierre Vernier (1580–1637). It is made to slide along the edge of any graduated instrument or regular scale, to indicate parts of divisions on the original scale.

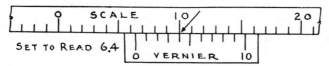

There are four terms defined in the above list which need further description—the meridian, the declination and reclination of a plane, and the vernier.

How to Find Your Meridian Line

It is just as necessary to place a dial properly as it is to be particular in laying out the hour lines, for if care is not taken in placing the dial, regardless of the care taken in inscribing the lines, it will be of little use.

The twelve o'clock line on a sundial should always lie in the plane of the meridian. Therefore the meridian line of the place must be determined in one way or another. In the case of certain dials, such as vertical declining dials, the meridian line must be determined before the hour lines can be constructed.

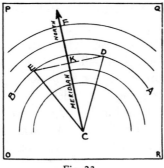

Fig. 23.

It is much easier to work in daylight than at night; consequently the following methods of finding the meridian have been selected because they depend upon the sun.

In Figure 23, the square OPQR represents a carefully leveled board. At any convenient place on the board mark the point C. With C as a center, describe several concentric circles. At C erect a pin perpendicular to the board and long enough to cast a shadow on the circles. Some time during the morning the shadow cast by the top of the pin will touch one of these circles, as on circle AB, at E. Mark this point carefully with a pencil. In the afternoon the shadow of the top of the pin will again touch the circle AB, at D. Mark this point carefully as before. Draw the line ED and find its middle point at K. From C, through K, draw the line CF, which will be the true meridian for the place.

Another convenient method, and one that consumes a minimum amount of time to accomplish, is by using the chart, Figure 22, which shows the relation between apparent noon (at which time the sun is on the meridian of the place every day) and mean time (the time shown by the clock). From the chart, find the time at which the sun will be on the meridian at any given place and on any particular day. At that time the shadow cast by a plumb line on a flat level surface will show the true meridian for that place.

This is easily accomplished by setting up a level board, or the surface on which the dial is to be placed may be used, Figure 24. Rig up a support from which a weight on the end of a string may be suspended. The weight is used to keep the string taut, and it should not touch or rest upon the board, neither should it be allowed to swing; therefore, unless it hangs free and motionless, it must be protected from the wind. Prior to the time the sun will be on the meridian as found from the chart, check your watch with WWV or CHU shortwave time signals. The telephone company is also a good source of time information.

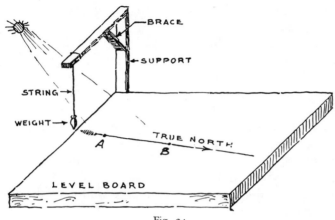

Fig. 24.

With your watch corrected and pencil in hand, go to the board a few minutes before the sun will arrive on your me-- ridian, to make sure that everything is all set. Then, at the proper time make two marks along the center of the shadow cast on the board by the string, as at A and B in the diagram. This done, a straight line drawn between the two points will lie in your meridian. The direction of north is obvious, for the shadow falls toward the north.

This is one of the easiest and most accurate methods of finding the meridian, without recourse to actual observation of the north star with an instrument, and saves many laborious hours of waiting and calculation.

How to Find the Declination of a Plane

Occasion may arise to place a dial on a surface that does not face the cardinal points of the compass. Before the hour lines for such a dial can be laid out it is necessary to know at what angle the plane, upon which the dial is to be drawn, declines from the meridian.

The accuracy of such a dial depends upon the care used in determining this angle of declination. In Figure 25, AB represents the side of a wall upon which it is desired to place a vertical dial. A board, OPQR, is pressed firmly against AB and leveled carefully. By one of the foregoing methods find the meridian line NS. Then draw the line EW, parallel to AB, cutting NS at D. From D draw the line DC, perpendicular to EW. The angle CDS is the declination of the plane upon which the dial is to be placed, and is also the declination of the dial. In the diagram, the wall *faces* the *south* and *declines east.*

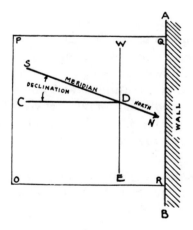

Fig. 25.

How to Find the Reclination of a Plane

A reclining dial depends for its accuracy upon the care with which the angle of reclination is found. Procure a flat board that has at least one good straight edge. Attach a piece of paper to the board, and lay the straight edge on the sloping wall or roof, as shown in Figure 26. The board should be perpendicular to the roof so that a flat weight suspended by a thin string will swing freely from it and just barely clear the paper. The use of the flat weight allows the string to lie nearly against the paper. When the weight has become motionless, take a pencil and mark two points on the paper on the line of the string as at E and F. Connect the two points thus found with a straight line, then draw a line, VW, parallel to the straight edge of the board. This line cuts EF at X. The angle EXW is the reclination sought.

The paper may now be removed from the board and used in laying out the hour lines.

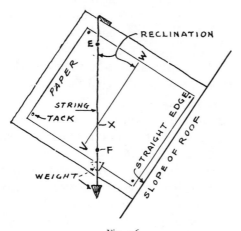

Fig. 26.

How to Set Your Dial

Any dial must be set carefully if it is to be at all serviceable. By setting, we mean placing the dial on some firm support that may be horizontal, vertical, reclining, or declining, and in the position for which it is made. In any case a meridian line must be found, in the manner previously described.

The easiest way to set a horizontal dial is, first mark the meridian line carefully on the top surface of the pedestal, or whatever surface is used to accommodate the dial; then extend the 12 o'clock line both ways to the edge of the dial plate and make faint marks at either extremity. Now place the dial on its pedestal so the two marks made on the edge coincide with the meridian previously marked on the pedestal. After the dial has been carefully levelled it will be in the position for which it was constructed.

In the case of vertical dials it is only necessary to place them in a vertical position so that the 12 o'clock line is perpendicular to the plane of the horizon, because the relation of the plane of the wall or roof with the horizon was determined before the lines were laid out. This can be done with a carpenter's level or the plumb line used to determine the meridian. The plumb line serves best to get the 12 o'clock line in the proper position. The dial face must line up with the wall or be parallel to it.

Where reclining and declining dials are concerned it is well to make sure the surface upon which the dial is to be fixed is firm and flat.

The horizontal dial can be set by finding the time the sun will be on the meridian of your place on a particular day from the chart, Figure 22. Carefully level the dial plate, as near its true position as possible. At the proper time turn the dial so that the time shown will be exactly 12 o'clock. This is

First (?) Dial in U. S. A.
Linoleum Pattern for Horizontal Dial.

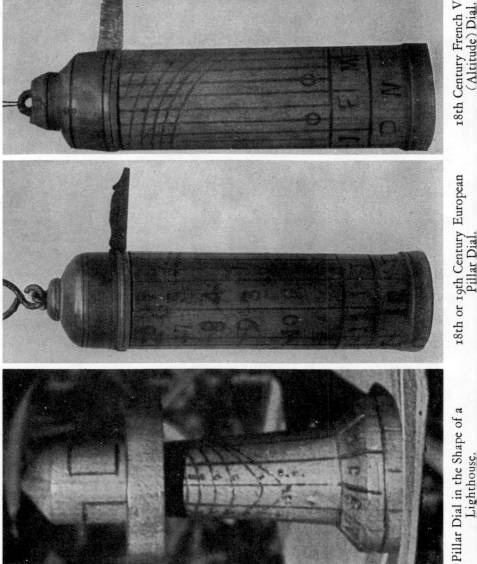

Pillar Dial in the Shape of a Lighthouse.

18th or 19th Century European Pillar Dial.

18th Century French Vertical (Altitude) Dial.

a quick easy method, but not so accurate as that described above.

How to Make a Vernier

The vernier is a small device, which enables one to determine more accurately fractional parts of the smallest division on a scale, than can be done by eye estimation. It is used on instruments such as transits, telescopes, and heliochrometers, where it is necessary to obtain a great degree of accuracy, either in setting or reading them. It can also be adapted to other dials where short spaces of time on the dial plate would be so small they could not be easily observed.

The principle of the vernier is simple. For instance, take any scale, or mark off any number of equal spaces as shown in Figure 27. If we wish to record tenth parts of the divisions on the scale, mark off on the vernier a distance equal to nine divisions on the scale. Divide this distance into ten equal parts; then each space on the vernier will be equal to 9/10 of a space on the scale, or 1/10 shorter. This means that the distance between the first mark on the vernier and the first mark on the scale, starting at 0, will be equal to 1/10 of a space on the scale; the distance between the next two marks

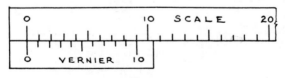

Fig. 27.

will be equal to 2/10 of a space on the scale and so on. Consequently, if the zero point of the vernier is shifted to the right 1/10 of a space on the scale, the first mark on the vernier will be directly opposite the first mark on the scale. If the zero point is again moved to the right 1/10 of a space on the scale the second mark on the vernier will be opposite the second mark on the scale, and so on.

To read a scale fitted with a vernier, Figure 28, first notice the position of the zero point or indicator of the vernier. In this case it lies between the fifth and sixth marks on the scale. Thus the first figure would be 5. Then look at the marks on the vernier and find out which one is directly opposite a mark on the scale—here, the fourth mark on the vernier is opposite the ninth mark on the scale. This means that the zero point of the vernier is 4/10 of the distance between the fifth and sixth marks on the scale. Thus the next figure will be 4, and the reading of the scale is five and four tenths (5.4).

It will be readily seen that only one mark on the vernier can be opposite a mark on the scale at one time. In order to apply a vernier to a sundial the hour lines must cut equal spaces on the edge of the dial. The equatorial dial is admirably suited to the use of a vernier.

Figure 29 shows the edge of an equatorial dial plate divided into five minute intervals. Two dials having the same divisions have been combined. On the left side, at A, we

Fig. 28.

have shown a vernier that will give the time to the nearest thirty seconds, and on the right side, at B, a vernier gives the time to the nearest minute. Without a vernier, the dial only shows positive readings to the nearest five minutes, for to obtain time any nearer than that, the position of the shadow would have to be guessed at or estimated.

In order to use the vernier, it must be attached to the dial plate so that it can be moved freely and remain wherever set either by friction or clamping, its inner edge riding against the edge of the dial. The vernier can also be placed on the dial surface if desired.

To obtain time by the vernier, set its zero or indicator line on the center of the narrow shadow as shown in the diagram, where at B the shadow lies between 11^h 15^m and 11^h 20^m in the morning. Then we look for a line on the vernier which is directly opposite a line on the dial; in this case, the second line on the vernier is opposite the line corresponding to 11^h 25^m. Therefore the shadow is located 2 minutes beyond the 15^m line, and shows the time to be 11:17 a.m.

The vernier at A has been set to show the time at 12: 21: 30 p.m. or 21^m 30^s after noon.

The rule by which the vernier is constructed can be summed up as follows:—

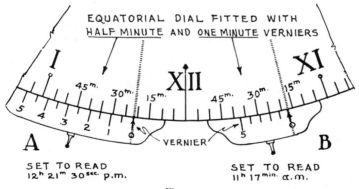

SET TO READ
12^h 21^m $30^{sec.}$ p.m.

SET TO READ
11^h $17^{min.}$ a.m.

Fig. 29.

Determine the number of spaces or divisions required on the vernier to give the desired parts of the smallest division on the scale. Then the length of the vernier is equal to the distance between the same number of divisions on the scale *less one*. In other words, if the vernier is to be divided into ten spaces, the length of the vernier is equal to nine of the smallest divisions on the scale. Similarly, at A in the diagram we wanted a vernier to read to 30 seconds. Because the small divisions of the dial are at five minute intervals, the vernier must have double the divisions that it would have if it were to record only to one minute, or ten spaces. Therefore, the length of the vernier is equal to the distance between nine five-minute lines on the dial. Then the vernier is divided into ten parts and labelled as shown.

Sundials become accurate instruments when equipped with verniers, and properly set up. Become familiar with your dial—know how it is constructed—know its accuracy—know how to use it.

VI

TIME AND STANDARD TIME DIALS

T HE reasons why a sundial tells time and why various kinds of time can be obtained from it were set forth in Chapter II. We are so dependent upon time, it is taken as a matter of course and is the thing, perhaps above all others, we know least about.

For those who wish to adapt their present dials to standard time or make dials which show only standard time, the following supplementary résumé of the means by which our present day time is obtained will help make the underlying principle of their construction more easily understood.

Timekeeping is based upon the period in which the earth makes one complete revolution upon its axis. For centuries this period, which we call the day, has been divided into twenty-four equal parts called hours, each hour consisting of sixty minutes and each minute of sixty seconds. This period may be measured by observing the daily motion of the stars or the sun. The period is determined when the object observed, having completed one revolution, returns to its starting point. For convenience, let us use the observer's meridian as a starting point. Then a solar day would be the interval between two successive crossings of that meridian by the sun.

Because the sun is not fixed as the stars appear to be, but moves irregularly in a path across the sky, completing a circuit in a year, it is only natural that this interval should vary. Obviously a clock which ticked off twenty-four hours in the interval of a solar day would be an irregular mechanism, difficult to make.

We look upon a good clock as one that runs uniformly, day after day. For such a clock the day must be uniform in length. If we average the values for the length of the solar days throughout the year a mean (average) solar day will be obtained. Therefore, if a clock is adjusted to tick off twenty-four hours during this mean interval, it will show what we call *mean solar time*. The irregular solar time is called *apparent* (*real*) *solar time*, in order to distinguish it from mean solar time.

Thus, if we had two clocks (one showing mean time and the other apparent time) reading 0 hours, and if we started them off simultaneously, it is evident that in a short time they would disagree. The difference between their readings is called the *equation of time*. The discrepancy between the two clocks is shown in the accompanying table for every day in the year, where the

Equation of Time=Mean Time—Apparent Time.

Therefore, it is possible to obtain the reading of one clock from the reading of the other for any day in the year, through the medium of the equation of time. Consequently, from the reading of a sundial, which shows apparent time, it is possible to obtain the mean time of the place or locality in which the dial is situated.

Too much importance cannot be placed upon the determination of the locality of a dial, because a sundial reads noon when the sun is on the meridian of the locality. The

Table Showing Equation of Time for Each Day in the Year

Day	Jan. min.	Feb. min.	Mar. min.	Apr. min.	May min.	June min.	July min.	Aug. min.	Sep. min.	Oct. min.	Nov. min.	Dec. min.
1	+ 3.6	+ 13.7	+ 12.5	+ 4.0	− 2.9	− 2.4	+ 3.6	+ 6.2	+ 0.0	− 10.2	− 16.3	− 11.0
2	4.0	13.8	12.3	3.7	3.1	2.3	3.8	6.1	− 0.3	10.5	16.4	10.6
3	4.5	13.9	12.1	3.4	3.2	2.1	4.0	6.1	0.6	10.9	16.4	10.2
4	5.0	14.0	11.9	3.1	3.3	1.9	4.1	6.0	0.9	11.2	16.4	9.8
5	5.4	14.1	11.7	2.8	3.4	1.8	4.3	5.9	1.2	11.5	16.3	9.4
6	+ 5.9	+ 14.2	+ 11.4	+ 2.5	− 3.5	− 1.6	+ 4.5	+ 5.8	− 1.6	− 11.8	− 16.3	− 9.0
7	6.3	14.3	11.2	2.2	3.5	1.4	4.7	5.7	1.9	12.0	16.3	8.6
8	6.7	14.3	11.0	2.0	3.6	1.2	4.8	5.6	2.2	12.3	16.2	8.1
9	7.1	14.3	10.7	1.7	3.7	1.0	5.0	5.4	2.6	12.6	16.1	7.7
10	7.6	14.4	10.5	1.4	3.7	0.8	5.1	5.3	2.9	12.9	16.0	7.2
11	+ 8.0	+ 14.4	+ 10.2	+ 1.1	− 3.7	− 0.6	+ 5.3	+ 5.1	− 3.3	− 13.1	− 15.9	− 6.8
12	8.4	14.4	9.9	0.9	3.8	0.4	5.4	5.0	3.6	13.4	15.8	6.3
13	8.7	14.4	9.7	0.6	3.8	0.2	5.5	4.8	4.0	13.6	15.7	5.9
14	9.1	14.3	9.4	0.4	3.8	0.0	5.6	4.6	4.3	13.9	15.5	5.4
15	9.5	14.3	9.1	0.1	3.8	+ 0.2	5.8	4.4	4.7	14.1	15.4	4.9
16	+ 9.8	+ 14.2	+ 8.8	− 0.1	− 3.8	+ 0.4	+ 5.9	+ 4.3	− 5.0	− 14.3	− 15.2	− 4.4
17	10.2	14.2	8.5	0.4	3.7	0.6	6.0	4.0	5.4	14.5	15.0	3.9
18	10.5	14.1	8.3	0.6	3.7	0.9	6.0	3.8	5.7	14.7	14.8	3.4
19	10.8	14.0	8.0	0.8	3.7	1.1	6.1	3.6	6.1	14.9	14.6	3.0
20	11.1	13.9	7.7	1.0	3.6	1.3	6.2	3.4	6.5	15.1	14.4	2.5
21	+ 11.4	+ 13.8	+ 7.4	− 1.2	− 3.6	+ 1.5	+ 6.2	+ 3.1	− 6.8	− 15.3	− 14.1	− 2.0
22	11.7	13.7	7.1	1.4	3.5	1.7	6.3	2.9	7.2	15.4	13.9	1.5
23	11.9	13.5	6.8	1.6	3.4	1.9	6.3	2.6	7.5	15.6	13.6	1.0
24	12.2	13.4	6.5	1.8	3.4	2.2	6.3	2.4	7.9	15.7	13.3	0.5
25	12.4	13.2	6.2	2.0	3.3	2.4	6.4	2.1	8.2	15.8	13.0	0.0
26	+ 12.6	+ 13.1	+ 5.8	− 2.2	− 3.2	+ 2.6	+ 6.4	+ 1.8	− 8.6	− 15.9	− 12.7	+ 0.5
27	12.9	12.9	5.5	2.4	3.1	2.8	6.4	1.5	8.9	16.0	12.4	1.0
28	13.0	12.7	5.2	2.5	2.9	3.0	6.3	1.3	9.2	16.1	12.1	1.5
29	13.2	.....	4.9	2.7	2.8	3.2	6.3	1.0	9.6	16.2	11.7	2.0
30	13.4	.....	4.6	2.8	2.7	3.4	6.3	0.7	9.9	16.3	11.4	2.5
31	+ 13.6	.....	+ 4.3	.....	− 2.6	.....	+ 6.3	+ 0.4	.....	− 16.3	.....	+ 3.0

E = M − A

Compiled from American Ephemeris

reading of a dial in one locality will differ from that of a dial in another locality east or west of it, by an amount equal to the difference in longitude of the two localities expressed in time. (See conversion table, Appendix II. Each degree of longitude is equal to four minutes of time.)

The watch time of everyday life is called *standard time,* because the time of all places in certain zones is referred to one meridian near the center of each zone, which is called the *standard time meridian.* In the United States we use the meridian having a longitude 75° west of Greenwich for Eastern Standard Time; 90° west for Central Standard Time; 105° west for Mountain Standard Time; and 120° west for Pacific Standard Time. Therefore, the conversion of the reading of a sundial in any locality to standard time requires:

(1) The reduction of the dial reading to the mean time of the locality, by the application of the equation of time.

(2) A reduction of the mean time of the locality to standard time, by the difference in longitude between the locality and the standard time meridian. This difference must be subtracted if the locality is east of the standard time meridian, and added if west.

The formula for finding the correction to be applied to any dial is:

Correction=Equation of Time + or — (Difference in Longitude × 4).

The following table shows the correction, to the nearest minute, which is to be applied to dials situated in longitude 78°, 75°, and 72°W, for a portion of the year, as found by the foregoing formula.

Month and Day	Correction for 78°	Correction for 75° (Stand. Time Merid.) = Eq. of Time	Correction for 72°
Feb.	min.	min.	min.
10	+26	+14	+ 2
15	26	14	2
20	26	14	2
25	25	13	1
Mar.			
1	+25	+13	+ 1
5	24	12	0
10	23	11	− 1
15	21	9	3
20	20	8	4
25	18	6	6
Apr.			
1	+16	+ 4	− 8
5	15	3	9
10	13	1	11
15	12	0	12
20	11	1	13
25	10	2	14
May			
1	+ 9	− 3	−15 etc.

It is evident that the formula for finding standard time from a dial is:

Standard Time=Apparent (dial) Time + or − the Correction.

Assume a dial on each of the meridians used in the preceeding example, which reads 3h 30m P.M., on March 20. Then, by using the corrections tabulated above, for those meridians, the standard time for the

78th meridian will be 3h 30m + 20m, or 3h 50m P.M.
75th meridian will be 3h 30m + 8m, or 3h 38m P.M.
72nd meridian will be 3h 30m − 4m, or 3h 26m P.M.

If proper attention is paid to the + and — signs preceding the figures in the tables and in the formulas, you should have no trouble computing the correction to be applied to a dial in any particular place, or in converting the dial reading to standard time.

The correction may be placed upon the dial plate in various ways; for example, in tabular form, arranging the figures around the dial, or incorporating it in the hour lines. Another method, which has been used on large dials, is that showing it in chart form, similar to Figure 22, Chapter V.

STANDARD TIME DIALS

You are now equipped with the necessary information for obtaining standard time, by the use of a table or chart, which will give the number of minutes to be subtracted from or added to the reading of an ordinary dial, any day in the year. Such charts and tables might well be confusing to your guests who are not so interested in sundials. A true standard time dial should not require any mental arithmetic, for all corrections are taken care of by mechanical or other means. All other dials are really just adapted to standard time.

We will show how an equatorial dial can be a standard time dial. That means, we are going to make a dial which will tell standard time, without any mental corrections. The illustration (B) opposite page 95 shows the general appearance of such a dial.

Before we go into the actual construction, let's see just how it works. The dial plate is movable and has to be set for each day. The settings are shown in the table on the dial plate. A fixed scale is placed at the bottom of the dial plate, and it is numbered to correspond to the setting figures in

the table. To set the dial we look at the table and find the
number opposite any particular day; then the 12 o'clock
line on the plate is brought opposite the corresponding
number on the scale; thus the dial will be properly set to
tell standard time all day long.

This is due to the fact the scale is so placed that the equation of time and difference in longitude have been accounted
for. The reasoning behind this construction is as follows.

The dial is figured for a standard time meridian. Furthermore, it was designed to be used by a person who did not
know anything about a dial and cared less. He just wanted
a dial to tell standard time; therefore no positive and negative quantities could be shown upon the dial plate or scale.
Since it is constructed for a standard time meridian, both
positive and negative corrections would have to be used, but
this was overcome by adding twenty minutes to the equation of time. Therefore the 12 o'clock line would be set at
20 on the scale when the equation of time is zero. The table
is, then, nothing more than the equation of time with twenty
added to it, thus giving the proper position for the 12 o'clock
line when it is set opposite a figure on the scale corresponding to the setting indicated in the table, eliminating negative
quantities.

If the dial is not to be used on a standard time meridian,
the scale must be shifted to the east or west by an amount
equal to the difference in longitude between the place and
its standard meridian. If your longitude is east, shift the
scale toward the east, if you are west shift the scale toward
the west. The setting table and figures on the scale will remain the same as for a standard time meridian.

The illustration below shows in detail a scale on a movable dial plate, which represents the equatorial ring of an

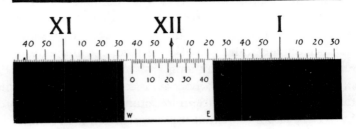

armillary, similar to that used on the heliochronometer in the illustration opposite page 126.

Another interesting variation of this type is that in the illustration (A) opposite page 95 where half a cylinder forms the dial plate, which can be moved by pressing on either the left or right hand edge; at the bottom of the plate is a setting scale. In the morning the right hand edge casts the shadow; in the afternoon the time is obtained from the position of the shadow cast by the left hand edge.

The Heliochronometer

Heliochronometers are made in many sizes and forms. Some make use of the conventional shadow, whereas others use a beam or ray of light which may either pass through a very narrow slit or be focused with a lens. We can find no better illustration of this finest and most accurate of sundials than the instrument pictured opposite page 126 and described many years ago in the popular German magazine "Die Himmelswelt".

The apparatus is heavy, accurately machined, and adjustable for various latitudes. The hours are marked on the inner surface of a ring which is held in place by a heavy brace. The housing to the right of the dial rim contains the days of the year on a wheel which is geared in proportion to the hour circle for setting; a second scale shows the longitude from Greenwich; and still another gives the difference from the Middle European standard time meridian. The base is fitted with two very sensitive levels, and a fine wire is used to cast the shadow. With such an instrument standard time can be obtained to within a few seconds, the accuracy depending upon the hour divisions on the dial and those on the vernier.

The heliochronometer is the acme of sundials. Its precision as to timekeeping and beautiful machining warrants as much admiration as a fine clock or watch. We hope the information given here will pave the way to making more standard time dials.

Time is like everything else in this world—it, too, has its humorous side. One day while we were looking at a very beautifully made dial which had a plate attached to it showing the correction to be added to the reading of the dial and thus obtain standard time, two ladies walked up to the dial. They seemed to be much interested in it. They looked at the sun—they looked at the dial—and they looked at their watches. After much looking, one turned to the other and said—"Isn't it too bad, Ella, that such a beautiful dial had to have all those corrections?" Whereupon Ella replied—"Someone certainly must have made a bad mistake in laying it out!" One might infer that it is rather dangerous to include a table of corrections for fear your best efforts may be ill considered, which is just another good reason to make standard time dials.

DAYLIGHT SAVING TIME

We have often heard of people twisting their dials to make them show Daylight Saving time. If this is done with a horizontal dial, the principle upon which it tells time is upset and can therefore never give you Daylight Saving time. If you must have your horizontal dial show Daylight time through the summer, paste new numerals over the old, then 12 o'clock would read 1 o'clock, 1 o'clock 2 o'clock, and so forth. Never twist the dial around.

The equatorial standard time dial mentioned above is ideally suited to the Daylight Saving period, because its dial plate moves or revolves about the gnomon without disturbing its direction. Therefore it is only necessary to use the 1 o'clock line instead of the 12 o'clock line for setting the dial. Any similar dial will be just as easily adjusted to Daylight Saving time.

At Riverside, California, a Concave Dial Face and
pierced Gnomon are employed.

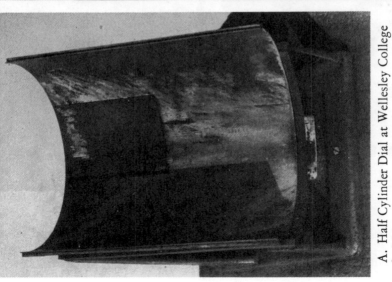

A. Half Cylinder Dial at Wellesley College in Massachusetts.

B. By using the Table on the Dial Face, Standard Time can be Read.

VII

HOW TO LAY OUT THE HOUR LINES

T HE following pages tell how to lay out the hour lines for many kinds of dials. It would be impossible to show the construction of all kinds and fortunately it is not necessary to do so. The kinds shown have been selected with care in order to provide the information for making as many types of dials as possible by adapting the principles of one to the other. For example it is easy to make a cross dial by applying the principle of the armillary or equatorial.

We have presented all diagrams and their accompanying descriptions in a standard form, to make the material easier to use. Each diagram shows, 1, the construction of the hour lines; 2, how the dial plate should look when completed; 3, the proper position for the dial when in use.

The supplementary text for each diagram gives first the important parts of each dial that must be known or found, before the hour lines can be laid out. Then follows the construction of the lines, the hour limitations or the number of hour lines necessary to be shown, and lastly how to set the dial.

In some instances, such as for the declining dial, a general note has been added at the end whenever necessary to make the work easier or as a guide in making other dials of similar character.

The names for each kind of dial are given in accordance with general usage.

We start by showing the construction of one of the simplest forms of all and the easiest to construct —the equatorial dial.

THE EQUATORIAL DIAL

PLATE I

The equatorial dial is one of the simplest forms of the sundial and one of the easiest to construct.

The plane of this dial lies parallel to the plane of the equator and it can be used at any place on the earth, provided the style is inclined at an angle above the horizon equal to the latitude of the place in which it is to be used.

The STYLE is a round rod, which passes through the dial plate and is perpendicular to it. It points to the celestial pole.

The SUBSTYLE is the center of the dial, at point E, Fig. 1.

The HEIGHT OF THE STYLE is determined by the size of the dial plate, and is usually from 6 to 8 inches high.

The Construction—Fig. 1

With E as a center, describe a small circle whose diameter is equal to the diameter of the style.

Also, with E as a center, describe the circle ABCD.

Draw EB, for the meridian; then draw the line 12a for the 12 o'clock line, parallel to EB and tangent to the small circle representing the diameter of the style.

THE EQUATORIAL DIAL

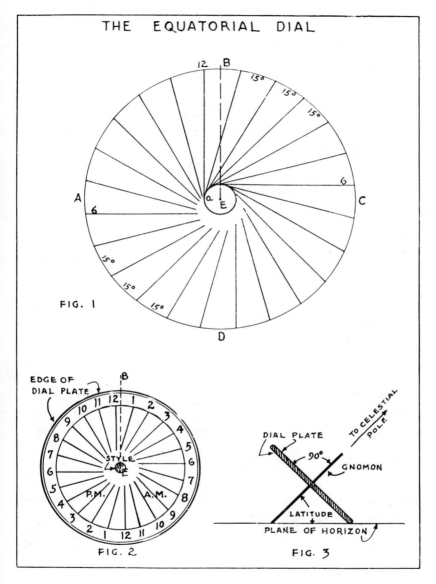

FIG. 1

FIG. 2

FIG. 3

Plate I.

Divide the circle ABCD into 24 equal parts, beginning at the point 12; and from the points thus found draw lines tangent to the small circle and on the same side with 12a. These lines will be the required hour lines.

When the style has been erected perpendicular to the dial plate the hour will be shown by the left-hand edge of the shadow.

Note:—If the style is less than 1/8" in diameter, or if the rod tapers to a point at the top, all the hour lines will be drawn from the center, at E; and the division of the hours will begin at the point B.

FIGURE 2 shows the hour lines transferred to the dial plate, and the method of numbering them on the upper or north face.

FIGURE 3 shows the position of the dial when in use.

Hour Limitations

This dial will show the time from sunrise to sunset throughout the year, if the hour lines are inscribed on both faces of the plate. Otherwise the dial will show only the time during the six months of summer, between the Equinoxes.

Setting the Dial

An equatorial dial must be so placed that the style points to the celestial pole, which will be at an angle above the horizon equal to the latitude of the place. The plane of the dial must be perpendicular to the style, and the 12 o'clock line must lie in the plane of the meridian.

The Horizontal Dial

PLATE II

The horizontal dial is the most common type of dial.

Its plane lies parallel to the plane of the horizon. The diagram shows the construction of the hour lines for latitude 43°10′N.

The STYLE points to the celestial pole.

The SUBSTYLE is the 12 o'clock line and lies in the plane of the meridian.

The HEIGHT OF THE STYLE is equal to the latitude of the place (43°10′ in the example).

The Construction—Fig. 1

Draw the horizontal line FAG (This will be the 6 o'clock line).

At A, draw AC perpendicular to FAG (This will be the 12 o'clock line).

Draw AD so that the angle DAC is equal to the latitude of the place (In this case 43°10′).

From B, on AC, draw BE perpendicular to DA.

Make BC equal to BE; then make AG and AF equal to AC.

Draw lines FC and CG. Through B draw a line parallel to FG, cutting CG at M, and FC at L. Through the points L and M draw the lines LK and MH parallel to AC.

With the radius BC, and centers at C, F, and G, draw the arcs TV, PQ, and SR. Divide these arcs into equal parts of 15° each. Draw lines from F, C, and G, through the points thus found, until they cut the lines KL, LM, and MH.

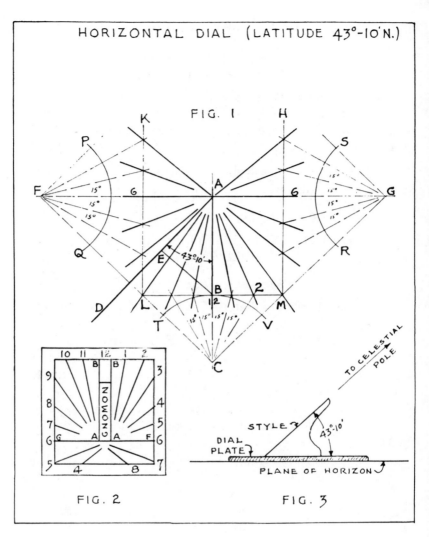

HORIZONTAL DIAL (LATITUDE 43°-10'N.)

FIG. 1

FIG. 2

FIG. 3

Plate II.

GAITHERSBURG, MARYLAND

This polar dial at the National Bureau of Standards is a memorial to Lyman James Briggs, the director from 1933 to 1946. The dial plate lies parallel to the earth's axis of rotation. In the middle of the dial are three gnomons, the upper and lower ones giving standard time, the middle one apparent time. For additional description of this dial, see pages 232 and 233.

Draw lines from A through the points found on KL, LM, and MH. Also draw lines from A through the points L and M. These will be the required hour lines.

The hours may be divided into halves, quarters, and so on, by further subdividing the arcs TV, PQ, and SR, into the desired number of parts.

FIGURE 2 shows the hour lines transferred to the dial plate, and the way in which they should be numbered.

FIGURE 3 shows the position of the dial when in use.

Hour Limitations

This dial will show the time from sunrise to sunset, in the latitude for which it is constructed, throughout the year.

Setting the Dial

To set the dial, first place it in position and carefully level it. Then turn it, so that the style points to the celestial pole and the 12 o'clock line lies in the plane of the meridian. See also Chapter V.

Note:—When the hour lines are transferred to the dial plate, allowance must be made for the width of the gnomon. This holds true for all dials. It has been exaggerated in Figure 2. Thus BB and AA represent the width of the style. Also note that the 7 and 8 hour lines in the evening do not converge in the same point as the afternoon hours, but on the opposite side of the gnomon where the morning hours converge, because they are the prolongation of the same hours in the morning. The same is true of the 4 and 5 hours in the morning.

The South Vertical Dial

PLATE III

The plane of the south vertical dial is perpendicular to the plane of the horizon, and faces due south. The construction of the hour lines for a dial in latitude 35°N is shown in the example.

The STYLE points to the celestial pole.

The SUBSTYLE is the 12 o'clock line and lies in the plane of the meridian.

The HEIGHT OF THE STYLE is equal to the complement of the latitude, which in this case is 55° (90° — 35° = 55°).

The Construction—Fig. 1

Draw the horizontal line FA (This will be the 6 o'clock line).

At A draw AC perpendicular to FA (This will be the 12 o'clock line).

Draw AD so that the angle CAD is equal to the height of the style, or 55°.

From B, on AC, draw BE perpendicular to AD.

Make BC equal to BE; then make AF equal to AC.

Draw the line FC. Through B draw a line parallel to FA, cutting FC at L. Through L draw the line LK parallel to AC.

With radius BC and centers at C and F, describe the arcs TV and PQ. Divide these arcs into equal parts of 15° each. Draw lines from F and C through the points thus found, until they cut the lines LK and LB, respectively.

Draw lines from A through the points found on LK and

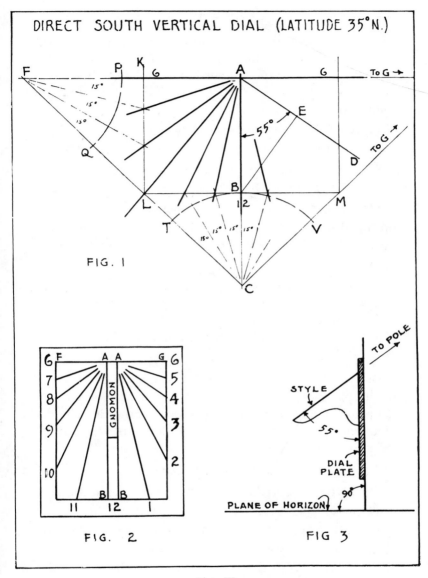

DIRECT SOUTH VERTICAL DIAL (LATITUDE 35°N.)

FIG. 1

FIG. 2

FIG 3

Plate III.

LB. Also draw a line from A through the point L. These lines will be those required for the morning hours.

To obtain the afternoon hour lines, extend the line FA to G, making AG equal to AC. Draw CG and continue the construction as shown above.

FIGURE 2 shows the hour lines transferred to the dial plate, and the way in which they should be numbered.

FIGURE 3 shows the position of the dial when in use.

Hour Limitations

The sun will not shine upon this dial before 6 in the morning or after 6 in the evening; therefore it is necessary to show only those hours between 6 a.m. and 6 p.m.

Setting the Dial

This dial must be placed in a perfectly vertical position, so that the 12 o'clock line will lie in the plane of the meridian, and the plane of the dial will face due south, or lie in the plane of the prime vertical. See also Chapter V.

THE NORTH VERTICAL DIAL
PLATE IV

The plane of the north vertical dial is perpendicular to the plane of the horizon and faces the true north. The diagram shows the construction of the hour lines for latitude 48°30′N.

The STYLE points to the celestial pole.

The SUBSTYLE is the 12 o'clock line and lies in the plane of the meridian.

The HEIGHT OF THE STYLE is equal to the complement of the latitude, which in this case is 41°30′ (90°— 48°30′ = 41°30′).

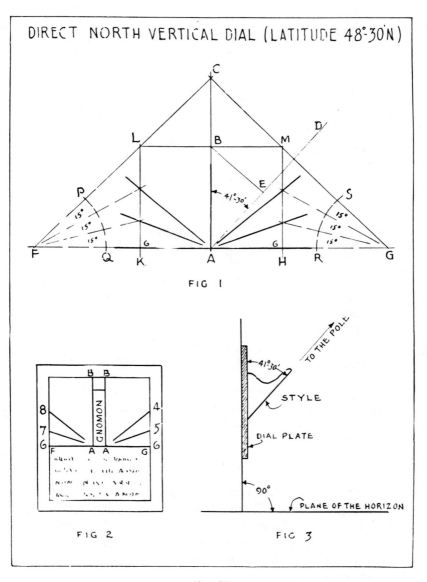

DIRECT NORTH VERTICAL DIAL (LATITUDE 48°30'N)

FIG 1

FIG 2

FIG 3

Plate IV.

The Construction—Fig. 1

Draw the horizontal line FAG (This will be the 6 o'clock line).

At A draw the line AC perpendicular to FAG (This will be the 12 o'clock line).

Draw AD so that the angle CAD is equal to the height of the style, or 41°30′.

From B, on AC, draw BE perpendicular to AD.

Make BC equal to BE; then make FA and AG each equal to AC.

Draw lines FC and CG. Through B draw a line parallel to FAG, cutting FC at L and CG at M. Through L and M draw lines LK and MH parallel to AC.

With radius BC and centers at F and G, describe the arcs PQ and SR. Divide these arcs into equal parts of 15° each. Draw lines from F and G through the points thus found, until they cut the lines LK and HM, respectively.

From A draw the required hour lines through the points found on LK and HM.

FIGURE 2 shows the hour lines transferred to the dial plate and the way in which they are numbered.

FIGURE 3 shows the position of the dial when in use.

Hour Limitations

Except for a few months in summer, between the Equinoxes, the sun will not shine upon this dial between the hours of 6 in the morning and 6 at night. This dial is generally found only on pillar dials (see Chapter IV); therefore it would not be necessary to inscribe the hours between 6 a.m. and 6 p.m.

Setting the Dial

This dial must be placed in a perfectly vertical position, so that the substyle line lies in the plane of the meridian and the face of the dial looks to the true north point of the horizon.

THE DIRECT EAST AND WEST VERTICAL DIALS
PLATE V

The planes of the direct east and west vertical dials lie in the plane of the meridian, and for this reason they are sometimes referred to as meridian dials. The hour lines for each dial are calculated in exactly the same way, so it is necessary only to describe the construction of one of them. The diagram shows the construction of the hour lines for the east dial, in latitude 52°30′N.

The GNOMON is usually made in the form of a flat rectangular bar, or in the shape of a pin. It is perpendicular to the face of the dial.

The STYLE points to the celestial pole and is parallel to the dial plate.

The SUBSTYLE is the 6 a.m. line (the 6 p.m. line in the west dial) and points to the celestial pole.

The HEIGHT OF THE STYLE is measured in inches (or millimeters) and is determined by the size of the dial plate. It is usually from 2½ to 3 inches in height.

The HOUR LINES are parallel to the substyle.

The Construction—Fig. 1

Draw the horizontal line AC. This represents the plane of the horizon.

DIRECT EAST AND WEST VERTICAL DIALS (LAT. 52°-30'N.)

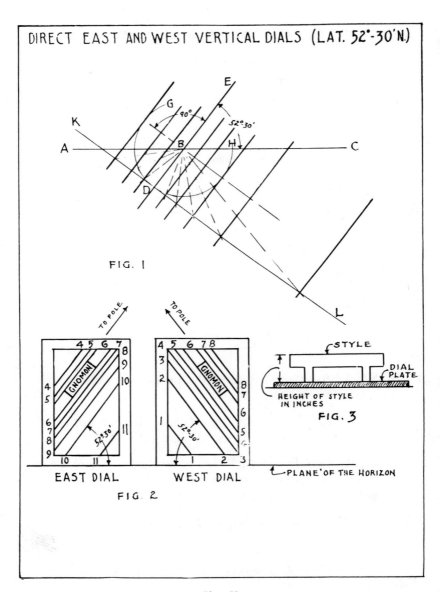

FIG. 1

TO POLE

TO POLE

EAST DIAL

WEST DIAL

FIG. 2

STYLE

DIAL PLATE

HEIGHT OF STYLE IN INCHES

FIG. 3

PLANE OF THE HORIZON

Plate V.

At B, on AC, draw DE so that the angle EBC is equal to the latitude of the place, in this case 52°30′. DE will also be the substyle line, and the 6 a.m. line.

Make BD equal to the desired height of the style in inches. Through D, draw the line KL perpendicular to DE.

With B as a center, and the radius BD, describe the arc GDH. Beginning with the point D, divide this arc into equal parts of 15° each, on each side of the line DE. From B draw lines through the points found on the arc until they cut the line KL.

Through the points thus found on KL draw lines parallel to DE, which will be the required hour lines.

FIGURE 2 shows the hour lines transferred to the dial plate, and the way in which they should be numbered. The west dial is also shown. Note the position of the lines on each face.

FIGURE 3 shows the gnomon generally used on this type of dial.

Hour Limitations

The east dial will show only the hours from sunrise to noon; the west dial, the hours from noon to sunset. These dials will not show the noon hour, because they lie in the plane of the meridian. The sun's rays at that time are parallel to the face of the dial; consequently the shadow cast by the gnomon will be infinite in length, and the edge of the shadow cannot be seen.

Setting the Dial

The plane of each dial must be perfectly vertical, and lie in the plane of the meridian.

The Polar Dial

PLATE VI

The plane of the polar dial is parallel to the axis of the earth, and if produced, would cut the celestial pole. The diagram shows the construction of the hour lines for any latitude.

The GNOMON is usually made in the form of a flat rectangular bar, or in the shape of a pin, and is perpendicular to the face of the dial.

The STYLE points to the celestial pole and is parallel to the face of the dial.

The SUBSTYLE is the 12 o'clock line and lies in the plane of the meridian.

The HEIGHT OF THE STYLE is measured in inches and is determined by the size of the dial. It is usually placed about $2\frac{1}{2}$ to 3 inches above the face of the dial.

The HOUR LINES are parallel to the substyle.

The Construction—Fig. 1

Draw the horizontal line AC.

At B, on AC, erect the perpendicular line BE (This will be the 12 o'clock line and the substyle).

Make BD equal to the desired height of the style.

With D as a center and the radius BD, describe the arc FBG. Beginning at B, divide this arc into equal parts of 15° each, on both sides of the line BD.

From D draw lines through the points found on arc FBG, until they cut the line AC. From the points thus found on

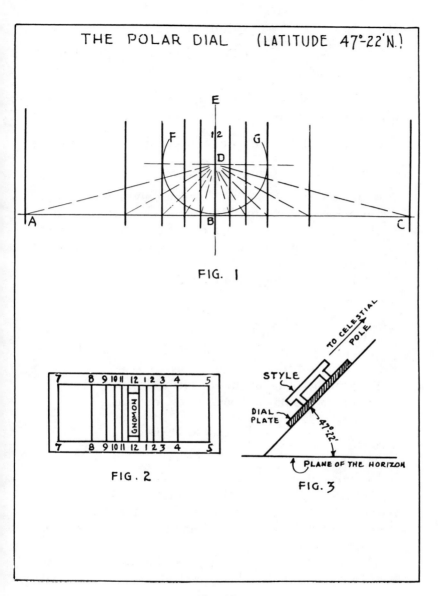

THE POLAR DIAL   (LATITUDE 47°-22'N.)

FIG. 1

FIG. 2

FIG. 3

Plate VI.

AC, draw lines parallel to BE, which will be the required hour lines.

FIGURE 2 shows the hour lines transferred to the dial plate, and the proper way of numbering them.

FIGURE 3 shows the position of the dial when in use.

Hour Limitations

It is necessary to show on this dial only those hours between 6 a.m. and 6 p.m., because the plane of the dial, if produced, would cut the east and west points of the horizon. At 6 in the morning and 6 in the afternoon the shadow cast by the gnomon is infinite in length; therefore the 6 a.m. and 6 p.m. lines cannot be placed upon this dial.

Setting the Dial

When setting the dial elevate the plate above the horizon, at an angle equal to the latitude of the place, as shown in Fig. 3; then turn the dial so that the 12 o'clock or substyle line lies in the plane of the meridian. If this is done correctly, the style will then point to the celestial pole.

THE DECLINING DIALS

PLATE VII

Declining dials are vertical dials so-called because they do not face the cardinal points of the compass. There are four types of declining dials: Those facing the south and declining toward the east or west; and those facing the north and declining toward the east or west.

The construction of each type is similar and only one need be described here. The example shows the construction of the hour lines for a south dial declining 28°W, in latitude 40°30′N.

The plane of this dial is perpendicular to the plane of the horizon, but it does not face any of the cardinal points. Unlike the preceding dials, two things must be known before the hour lines can be constructed:—first, the latitude of the place; and second, the declination of the dial or the declination of the plane upon which the dial is to be placed. (The declination of the plane may be found by one of the methods described in Chapter V).

The GNOMON is perpendicular to the face of the dial.

The STYLE points to the celestial pole.

The SUBSTYLE is to be determined (The substyle is not the 12 o'clock line, in this type of dial).

The HEIGHT OF THE STYLE is to be determined.

The Construction

Note:—For clarity, the construction diagram has been divided into two parts. In practice, Fig. 1 would be incorporated in Fig. 2.

To Find the Substyle Line—Fig. 1

Draw the horizontal line ABC.

From B let fall a perpendicular line BD, which will be the meridian or 12 o'clock line.

Draw the line BE so that the angle DBE is equal to the complement of the latitude, which in this case is 49°30′ (90° — 40°30′ = 49°30′).

With B as a center and any convenient radius, draw the

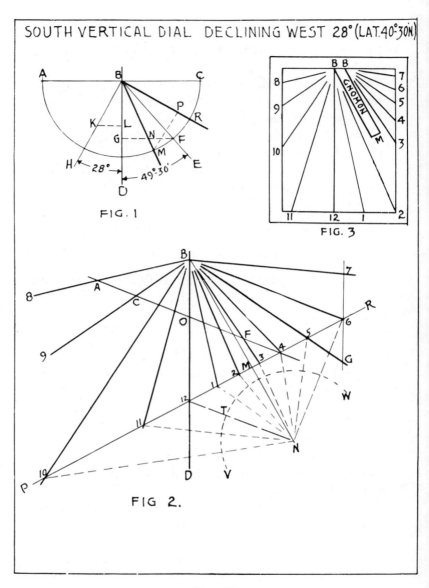

SOUTH VERTICAL DIAL   DECLINING WEST 28° (LAT.40°30'N)

FIG. 1

FIG. 3

FIG 2.

Plate VII

arc AC, which cuts the line BE at F. From F draw a line perpendicular to BD at G.

From B draw BH, making the angle HBD equal to the declination of the dial, which in this case is 28°.

Make BK equal to GF, and from K draw KL perpendicular to BD. Then on GF make GN equal to KL.

Draw a line from B through N, cutting the arc AC at M. This line is the substyle line upon which the gnomon must be erected, perpendicular to the face of the dial.

It will be noticed that the substyle line must fall among the afternoon hours if the dial declines west; among the morning hours if the dial declines east.

To Find the Height of the Style—Fig. 1

With N as a center and the radius BL, describe an arc cutting the arc AC at R.

Draw a line from B through the point R.

The line BR is the style, and the angle RBM is the height of the style. (The style must make an angle with the face of the dial equal to the angle RBM).

To Find the Hour Lines—Fig. 2

In this figure the lines BD and BM have been reproduced from Figure 1. In Fig. 1 the line MP is perpendicular to BR.

In Figure 2 the line MB has been produced so that MN is equal to MP (Fig. 1).

With N as a center and any convenient radius, describe the arc VW; and at M draw the line PR perpendicular to BM, cutting BD at 12.

Now draw N 12, cutting VW at T.

Beginning at T divide the arc VW into equal spaces of

15° each. Draw lines through these divisions until they cut
the line PR, at 10, 11, 12, and so on.

From B draw lines through the points 10, 11, 12, and so
on. These lines will be the required hour lines.

To obtain the hours after 6 p.m. draw a line through 6
(B6 is the 6 o'clock line) parallel to BD, and cutting B5 at
G. For the 7 p.m. line make the distance 67 equal to G6.
The 8 p.m. line may be obtained in the same way.

To obtain the hours before 10 a.m., draw a line through
any point on BD, such as O, parallel to B6 (the 6 o'clock
line). This line cuts B3 at F, and B4 at 4. From O lay off
the distance OC equal to OF; and OA equal to O4. Lines
drawn from B through the points C and A will give the
9 a.m. and 8 a.m. lines.

Figure 3 shows the hour lines applied to the dial plate,
in their proper position, and the way in which they should
be numbered.

Hour Limitations

The length of time the sun shines on this dial and the
number of hour lines to be inscribed is governed by the declination. A simple method of determining what hour lines
should appear on a declining dial, where great accuracy is
not desired, is by converting the declination from degrees to
hours and minutes. One hour of time is equal to 15°. Therefore in the above example, 28° is just a little less than two
hours. Thus, the sun would not shine on this dial until about
8 a.m. nor would the sun cease to shine on it until about
8 p.m. It must be remembered that this is not an accurate
method, for the latitude and time of year should be considered where extreme accuracy is desired.

The 12 o'clock line is always a vertical line.

Setting the Dial

In setting the dial, it is essential that the plane upon which it is to be placed is vertical. The declination of the plane having been carefully determined, attach the dial securely.

Declining Dials in General

In the foregoing example, the construction of the hour lines for a south dial declining 28°W, in latitude 40°30'N, has been shown in detail.

These lines will also serve for a south dial declining east, or a north dial declining west, and a north dial declining east, provided each dial has the same declination and latitude. This will be easily seen if the hour lines have been drawn on transparent paper.

Thus, if Fig. 3 is looked at from the rear, the hour lines for a south dial declining east 28° will be seen; if the figure is turned upside down, the hour lines will be those for a north dial declining west 28°; and the reverse side of the figure, when turned upside down, will show the hour lines for a north dial declining east 28°.

It must be remembered that the morning hours of a dial declining west will become the afternoon hours of a dial declining east; and that the substyle of a dial declining west will fall among the afternoon hours. But the substyle of a dial declining east will fall among the morning hours.

Therefore, while making one dial, the hour lines for all four declining dials have been constructed.

THE RECLINING DIALS

PLATE VIII

There are four types of reclining dials: the direct south, north, east, and west dials. These may be subdivided into two groups—the north-south and east-west.

Two reclining dials have been described in the foregoing pages: the equatorial, a north recliner; and the polar, a south recliner. These two dials take their names from the planes in which they lie. For this reason they are not usually classed with the reclining dials.

The direct reclining dials are so-called because their planes face the cardinal points of the compass, and as you stand before them they lean from you (or recline from the zenith).

A plumb line is perpendicular to the plane of the horizon and if extended, it would cut the zenith at any given place. The reclination of a dial or plane is that angle, measured in degrees, formed by the intersection of the dial or plane with a plumb line. See also Chapter V.

Before the hour lines can be computed the dials must be referred to that position in which they would become horizontal or vertical dials. This is called "reducing to a new latitude". The method of reduction is but a simple arithmetical operation.

The Construction of the North-South Reclining Dials

Direct north and south reclining dials must be reduced to *new latitudes,* where they will *become horizontal dials.* First, determine the reclination of the plane upon which the hour lines are to be inscribed, and then proceed as follows:

In the case of the south recliner—
If the reclination of the dial is *less* than the *complement* of the *latitude,* the NEW LATITUDE=the complement of the latitude *minus* the reclination.

If the reclination is *equal* to the *complement* of the *latitude,* the dial will be a *polar dial.*

If the reclination of the dial is *greater* than the *comple-*

ment of the *latitude,* the NEW LATITUDE=the reclination *minus* the complement of the latitude.

In the case of the north recliner—

If the reclination is *less* than the *latitude,* the NEW LATITUDE=the complement of the latitude *added* to the reclination.

If the reclination is *equal* to the *latitude,* the dial will be an *equatorial dial.*

If the reclination is *greater* than the *latitude,* the NEW LATITUDE=180° *minus* the (reclination *added* to the complement of the latitude).

From the above formulas, notice that, in each case,—

The STYLE points to the celestial pole.

The SUBSTYLE is the 12 o'clock line and lies in the plane of the meridian.

The HEIGHT OF THE STYLE is equal to the new latitude.

The construction of the hour lines is the same as that for a horizontal dial. (See page 99).

The Construction of the East-West Reclining Dials

The direct east and west reclining dials can be reduced to latitudes in which they will be south vertical declining dials. This new latitude may be found very simply by the following formula:

The *complement* of the *latitude* of the place is equal to the NEW LATITUDE, wherein the dial becomes a *south*

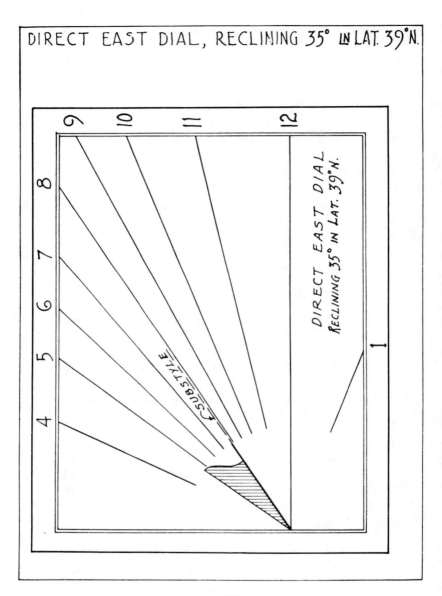

Plate VIII.

vertical declining dial; and the *complement* of the reclination is equal to the *declination* of the south vertical dial in
the *new latitude.*

Note:—An east recliner will have a west declination, in
the new latitude, and vice versa.

Having found the new latitude and the declination of
the dial in that latitude, proceed to lay out the hour lines for
a south vertical declining dial according to the construction
described on page 112.

EXAMPLE: Let it be required to construct an east dial
reclining 35°, in latitude 39°N.

From the formula given above, this reclining dial will become a vertical dial in latitude 51° (90°—39°=51°), and
decline in that latitude 55° west (90°—35°=55°).

Plate VIII shows the appearance of this dial, and the way
in which the hours should be numbered. The twelve o'clock
line is at the base of the dial and lies parallel to the plane of
the horizon. The center of the dial, on the east recliner, is
at the left; and at the right, on the west recliner.

If the hour lines are drawn on transparent paper, the reverse side will show the hour lines for a west dial reclining
35° in latitude 39°N.

If Plate VIII is turned so that the 12 o'clock line is perpendicular, we have a south vertical dial declining 55° W in
latitude 51°; the morning hours then become afternoon
hours.

Therefore, in each of these dials:

The GNOMON is perpendicular to the face of the dial.

The STYLE points to the celestial pole.

The SUBSTYLE is to be determined. (The substyle is not the 12 o'clock line in this type of dial.)

The HEIGHT OF THE STYLE is to be determined.

Setting the Reclining Dials

Care must be used to set the dial in the position for which it was computed. The 12 o'clock line will be near the bottom of the dial and must lie in the plane of the meridian and parallel to the plane of the horizon.

RECLINING-DECLINING DIALS
PLATE IX

We have had but one inquiry in the last twelve years about the construction of these dials, but the book would not be complete without them.

The Reclining-Declining dials neither face any of the cardinal points of the compass, nor do they stand upright. Three things must be known before the hour lines can be laid out—the height of the style, the substyle distance, and the angle which the meridian (12 o'clock line) makes with the horizon. (The declination and reclination may be found by one of the methods described in Chapter V.)

The GNOMON is perpendicular to the face of the dial.

The STYLE points to the celestial pole.

The SUBSTYLE is to be determined (The substyle is not the 12 o'clock line in this type of dial.)

The HEIGHT OF THE STYLE is to be determined.

The MERIDIAN LINE is to be determined (The meridian or 12 o'clock line makes an angle with the horizon in this type of dial.)

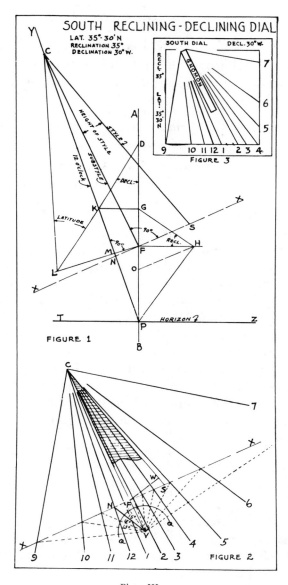

SOUTH RECLINING-DECLINING DIAL
LAT. 35°-30' N
RECLINATION 35°
DECLINATION 30° W.

FIGURE 1

FIGURE 2

FIGURE 3

SOUTH DIAL   DECL. 30° W.

Plate IX.

The Construction

Note: Although the work can be done on one diagram, the accompanying plate makes use of two figures for the sake of clarity.

To find the Meridian Line, and its Angle with the Horizon—Fig. 1

Draw the perpendicular line AB, and at any convenient point such as F, erect FH perpendicular to AB. Make FH any convenient distance or equal to the height of a perpendicular style.

At H draw HG, making the angle FHG equal to the reclination of the dial (if the dial *inclines,* leans toward you, the line HG is drawn downwards). Then draw HP so that angle FHP is equal to the complement of the reclination.

Draw GK perpendicular to AB and G; make GD equal to GH; then draw DK, making an angle with AB equal to the declination of the dial. (If the dial declines to the west, this angle is set off to the left of AB; if the dial declines east, the angle is set off to the right of AB).

Draw PY through K. This will be the meridian or 12 line of the dial. TZ is drawn, through P perpendicular to AB. Then, the angle TPK is the angle the meridian makes with the horizon.

To Find the Center of the Dial—Fig. 1

From F draw FL perpendicular to PY and cutting PY at M.

Make FO equal to FM; and make ML equal to OH. Then draw LC, which makes an angle with KL equal to the latitude of the place. The intersection of CL with PY at C is the center of the dial.

To Find the Substyle Line—Fig. 1

A line drawn from C to F will be the substyle line.

To Find the Height of the Style—Fig. 1

Through F draw XX perpendicular to the substyle CF. Make FS equal to FH; then draw CS. The line CS will be the style and the angle SCF is equal to the height of the style. The line XX cuts PY at N.

To Find the Hour Lines—Fig. 2

In the figure, the points C, S, F, N, and the line XX correspond to the same points in Figure 1.

In Figure 2, draw a line from F to CS, which will be perpendicular to CS at W.

Extend CF and make FV equal to FW. Connect V and N. With V as a center describe the arc QQ and beginning at the point where the line VN cuts the arc QQ, divide this arc into equal spaces of 15° each. Draw lines through these divisions until they intersect the line XX.

Then, lines drawn from C through the intersections on XX will be the required hour lines.

Figure 3 shows the way in which the hours should be numbered on a dial in lat. 35°30′N, recl. 35° and decl. 30°W.

Hour Limitations

The length of time the sun will shine on reclining-declining dials, depends upon the amount of their reclination and declination. See hour limitations for declining dials.

Setting the Dial

The reclination and declination of the plane upon which the dial is to be placed having been carefully found, the dial should be securely attached.

THE ARMILLARY SPHERE.

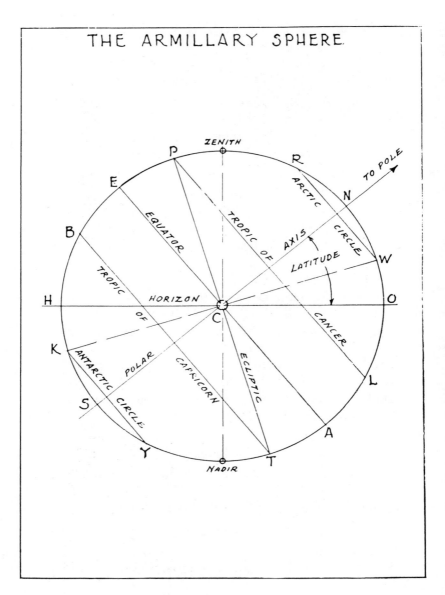

Plate X.

A German Heliochronometer.

Dial at Kitt Peak, Arizona.

Wall Dial on Library, Bowdoin College, Brunswick, Maine.
Instruments at Observatory Built by Emperor Kublai Khan, 1296 A. D., Peiping China.

The Armillary

PLATE X

The armillary consists of several rings put together in the form of a hollow sphere. Usually ten rings are employed, denoting the ten major circles of the celestial and terrestrial spheres placed in proper relation to each other. These circles represent the (1) Meridian, (2) Horizon, (3) Equator, (4) Ecliptic, (5) Tropic of Cancer, (6) Tropic of Capricorn, (7) North Polar or Arctic Circle, (8) South Polar or Antarctic Circle, (9) Equinoctial Colure, and (10) Solstitial Colure.

The Construction

The circle HENAS represents the meridian; the line SN the polar axis of the sphere, with its poles at N and S.

The point C marks the center of the sphere and the horizontal line HO, drawn through C, represents the horizon, which makes an angle with the polar axis SCN equal to the latitude of the place.

The Equator, EA, is perpendicular to the polar axis at C.

The Ecliptic, PT, is drawn through C (which also represents the east and west points of the horizon) making an angle with the equator equal to 23°27', cutting the sphere at P and T. The points P and T denote the greatest northern and southern declination of the sun.

The Tropic of Cancer is shown by PL, which is drawn through P parallel to the equator, because it is a circle of latitude.

The Tropic of Capricorn, TB, is drawn through T parallel to the equator.

The North and South Polar Circles are similarly drawn. The dotted line KCW is perpendicular to the ecliptic at C, and cuts the sphere at W and K. These two points are called the north and south poles of the ecliptic. The north and south polar circles are noted by the lines WR and KY, drawn through W and K, parallel to the equator.

The Equinoctial Colure is indicated by the line SCN. It passes through the points where the ecliptic crosses the equator, and the north and south poles of the sphere.

The plane of the Solstitial Colure is perpendicular to the plane of the equinoctial colure and it passes through the poles of the sphere, the zenith, and the nadir. In the diagram it coincides with the meridian, shown by the circle SENA.

The Armillary as a Sundial

The DIAL PLATE will be the inner surface of the ring representing the Equator.

The GNOMON is a thin round rod, extending through the sphere as SCN, and lies parallel to the axis of the earth.

The STYLE is essentially the gnomon, unless it is quite large in diameter. It lies parallel to the earth's axis and points to the celestial pole.

The SUBSTYLE is the 12 o'clock line.

The HEIGHT OF THE STYLE. In this case it is the angular distance of the style above the plane of the horizon, which is equal to the latitude of the place.

The hour lines are easily constructed. The method is the same as that for the equatorial dial shown on page 96.

Beginning at the point A, which represents the intersection of the equator and meridian, mark off equal spaces of 15° each on the dial plate. Thus will the morning hours be laid out on the west side of the sphere and afternoon hours on the east side.

The Signs of the Zodiac are often found on the armillary. They should be placed on the circle representing the ecliptic, beginning at the point where the ecliptic crosses the equator with the Vernal Equinox or Sign of Aries at the west. Thus will the sign of Cancer be at the point P, and the Sign of Capricornus will appear at the point A, and so on. For a more detailed account of the Signs see page 132.

Hour Limitations

This dial will show the time from sunrise to sunset throughout the year.

Setting the Dial

The twelve o'clock line must lie in the plane of the meridian and the axis or gnomon must be elevated above the plane of the horizon at an angle equal to the latitude of the place.

Note: Although the ten major circles of the sphere have been described, it is not necessary to use them all, if the instrument is to be used as a sundial; but it must be remembered that the hour lines are inscribed on the equatorial circle and the Signs of the Zodiac on the ecliptic.

VIII

DIAL FURNITURE

T HE essential lines on a sundial are those representing the hours of the day, with their accompanying figures. All other lines and symbols constitute the "furniture" of the dial.

The furniture most commonly found on dials shows: (1) the difference between apparent and mean time· (equation of time); (2) the sun's declination throughout the year; (3) the time of sunrise and sunset; (4) the Signs of the Zodiac and the date of the sun's entrance into each; and (5) the points of the compass. Other mathematical and astronomical data may be added, such as meridian lines that show when it is noon at any particular place; the Babylonian hours (reckoned from sunrise to sunset); the Jewish hours (the old, unequal planetary hours); and the Italian hours (beginning at sunset). Such lines increase the interest and usefulness of the dials displaying them.

There seemed to be no limit to the amount of furniture that early dialists were wont to place upon a single dial. Figure 30 is an admirable example of such a dial, made in the 17th century, which has more lines than most people would want to compute. Aside from the time of day the facts depicted on this dial are varied and interesting, therefore a short description of them will not be amiss. Upon it are drawn:

·I·COUNT·ONLY·SUNNY·HOURS·

VI   VI

VII   V

VIII   IV

IX   X XI XII I II III

·DISCE·VT·SEMPER·VICTVRVS·
·VIVE·VT·CRAS·MORITVRVS·

STOW, MASSACHUSETTS

The large size and clean design of this sundial on a schoolhouse make it an ideal instructional aid. The motto in Latin reads: *Learn as if you were going to live forever. Live as if you were going to die tomorrow.*

1—Lines of declination, which show the path of the shadow of the nodus when the sun is on the equator and in the two tropics. On the meridian or substyle line is marked the position of the shadow of the nodus for each degree of declination.

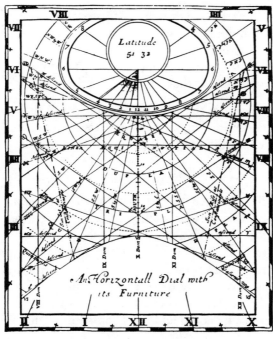

Fig. 30.

2—Azimuth lines, which show the position of the sun, throughout the day, with respect to the points of the compass; or its angular distance east and west of the meridian.

3—Lines showing the length of the day; the time of sunrise and sunset.

4—The dial is constructed for London, but the time of sunrise and sunset in Constantinople is also shown.

5—Lines showing the rising and setting of the Signs of the Zodiac (ascending and descending Signs); and the position of the sun with respect to the Signs. These lines were used by astrologers to tell the position of the sun in relation to its Cuspis; and they did not have any astronomical application.

6—Lines showing the altitude of the sun, its angular height above the horizon.

7—Date of the sun's entrance into each Sign of the Zodiac, and its declination at that time.

It is evident that the computation of such a dial would require a good knowledge of celestial mechanics; and also, in the 17th century, the services of an expert engraver.

THE LINES OF DECLINATION AND ZODIACAL SIGNS

Early dialists often placed upon their dials certain lines, called lines of declination, which recorded the entrance of the sun into the various Signs of the Zodiac. This gave them a measure of time, because it takes the sun about a month to pass from the beginning of one sign to the beginning of the next. Feast days, Holy days, events of importance, and the time of year were also shown by these lines; and if one wished to be facetious he could add lines commemorating birthdays, wedding anniversaries, and so on.

Today, such lines are usually used for ornamentation rather than for the utilitarian purposes of not many centuries ago. Since the location of these lines on the dial plate depends upon the position of the sun, they have not entirely lost their usefulness, even in this day and age. They have an educational value, for by them one may obtain a clearer

Day	Jan. °	Jan. ′	Feb. °	Feb. ′	Mar. °	Mar. ′	Apr. °	Apr. ′	May °	May ′	June °	June ′	July °	July ′	Aug. °	Aug. ′	Sep. °	Sep. ′	Oct. °	Oct. ′	Nov. °	Nov. ′	Dec. °	Dec. ′
1	−23	2	−17	9	−7	39	+4	28	+15	1	+22	2	+23	8	+18	5	+8	22	−3	6	−14	22	−21	47
2	22	56	16	52	7	16	4	51	15	19	22	10	23	4	17	50	8	0	3	29	14	41	21	56
3	22	51	16	34	6	53	5	14	15	37	22	18	22	59	17	34	7	38	3	53	15	0	22	5
4	22	45	16	17	6	31	5	37	15	54	22	25	22	54	17	19	7	16	4	16	15	19	22	13
5	22	38	15	59	6	7	6	0	16	12	22	32	22	49	17	3	6	54	4	39	15	37	22	21
6	−22	31	−15	40	−5	44	+6	23	+16	29	+22	38	+22	43	+16	46	+6	32	−5	2	−15	56	−22	29
7	22	24	15	22	5	21	6	45	16	46	22	44	22	37	16	30	6	9	5	25	16	14	22	36
8	22	16	15	3	4	58	7	8	17	2	22	50	22	31	16	13	5	47	5	48	16	31	22	42
9	22	8	14	44	4	34	7	30	17	18	22	55	22	24	15	56	5	24	6	11	16	49	22	49
10	21	59	14	24	4	11	7	52	17	34	23	0	22	16	15	38	5	1	6	34	17	6	22	54
11	−21	50	−14	5	−3	47	+8	15	+17	50	+23	5	+22	9	+15	21	+4	39	−6	57	−17	22	−22	59
12	21	41	13	45	3	24	8	37	18	5	23	9	22	1	15	3	4	16	7	19	17	39	23	4
13	21	31	13	25	3	0	8	59	18	20	23	12	21	52	14	45	3	53	7	42	17	55	23	9
14	21	20	13	5	2	36	9	20	18	35	23	16	21	43	14	26	3	30	8	4	18	11	23	13
15	21	10	12	44	2	13	9	42	18	49	23	19	21	34	14	8	3	7	8	27	18	26	23	16
16	−20	58	−12	24	−1	49	+10	3	+19	3	+23	21	+21	25	+13	49	+2	44	−8	49	−18	42	−23	19
17	20	47	12	3	1	25	10	24	19	17	23	23	21	15	13	30	2	21	9	11	18	57	23	21
18	20	35	11	42	1	2	10	45	19	31	23	25	21	4	13	11	1	57	9	33	19	11	23	24
19	20	23	11	20	0	38	11	6	19	44	23	26	20	54	12	51	1	34	9	54	19	25	23	25
20	20	10	10	59	−0	14	11	27	19	56	23	26	20	43	12	32	1	11	10	16	19	39	23	26
21	−19	57	−10	37	+0	10	+11	48	+20	9	+23	27	+20	31	+12	12	+0	47	−10	38	−19	53	−23	27
22	19	43	10	16	0	33	12	8	20	21	23	27	20	20	11	52	0	24	10	59	20	6	23	27
23	19	29	9	54	0	57	12	28	20	33	23	26	20	8	11	32	+0 / −0	1	11	20	20	19	23	27
24	19	15	9	32	1	21	12	48	20	44	23	26	19	55	11	11	−0	23	11	41	20	31	23	26
25	19	0	9	9	1	44	13	8	20	55	23	24	19	43	10	51	0	46	12	2	20	43	23	25
26	−18	46	−8	47	+2	8	+13	27	+21	6	+23	23	+19	30	+10	30	−1	9	−12	23	−20	55	−23	23
27	18	30	8	25	2	31	13	46	21	16	23	21	19	16	10	9	1	33	12	43	21	6	23	21
28	18	15	8	2	2	55	14	5	21	26	23	18	19	3	9	48	1	56	13	3	21	17	23	18
29	17	59	..	...	3	18	14	24	21	35	23	15	18	49	9	27	2	20	13	23	21	27	23	15
30	17	43	..	...	3	42	14	43	21	35	23	12	18	34	9	7	2	43	13	43	21	37	23	11
31	−17	26	..	...	+4	5			+21	53			+18	20	+8	44			−14	3			−23	7

Dates of Equinoxes and Solstices underscored. (Compiled from the American Ephemeris) conception of the motion of the earth in relation to its all important luminary—the sun.

The sun, in its apparent movement among the stars, traces out a path called the *ecliptic,* the plane of which is inclined to the plane of the celestial equator at an angle of about 23°27'. During one half of the year the sun appears north of the celestial equator and during the other half, south of it. The sun's distance north or south of the equator is called its *declination,* (expressed in degrees and minutes of arc), which varies from day to day. The amount of this declination for each day in the year, at apparent noon, is given in the accompanying table where the northern declination is preceded by a plus (+) sign and the southern declination by a minus (−) sign.

Although data have been omitted, which may be easily obtained from any good almanac, this table is inserted because it is not always found in a convenient form for use in the construction of the lines of declination.

These lines are also known as the Arcs of the Signs, and on early dials the zodiacal symbols were often placed at their extremities. The Zodiac is a zone in the sky 16° wide (8° on each side of the ecliptic), in or near which the planets and sun appear to move. Beginning at the point on the ecliptic, which marks the position of the sun at the Vernal Equinox, this zone or belt is divided into 12 parts of 30° each, called Signs. The Signs derive their names from the constellations with which they coincided, about 2000 years ago.

The Signs meant much to the ancients, who were well acquainted with the meanings and omens attached to each. Even today, the entrance of the sun into the Sign of Aries marks the beginning of spring; and summer begins when it enters the Sign of Cancer.

The following table shows the zodiacal symbol attendant to each Sign and the approximate date of the sun's entrance into each Sign.

Symbol	Name		Date of Sun's Entrance
♈ ....Aries.........Ram	} Spring Signs		⎧ March 21
♉ ....Taurus........Bull			⎨ April 20
♊ ....Gemini.......Twins			⎩ May 21
♋ ....Cancer........Crab	} Summer Signs		⎧ June 21
♌ ....Leo...........Lion			⎨ July 23
♍ ....Virgo.........Virgin			⎩ August 23
♎ ....Libra.........Balance	} Fall Signs		⎧ September 23
♏ ....Scorpius.......Scorpion			⎨ October 24
♐ ....Sagittarius....Archer			⎩ November 22
♑ ....Capricornus...Goat	} Winter Signs		⎧ December 22
♒ ....Aquarius......Water-Bearer			⎨ January 20
♓ ....Pisces.........Fishes			⎩ February 19

One must not lose sight of the fact that due to the precession (retrograde or backward motion) of the equinoxes along the ecliptic, each Sign has moved backward $30°$ into the constellation west of it; so that today, the Sign of Aries is in the constellation of Pisces, and so on. The Signs are independent and they have no connection with the apparent position of the sun in the constellations.

Thus the usefulness of a dial would be increased if the names of the zodiacal constellations are placed upon it as well as the attendant Signs. In order to do this, find the date upon which the sun enters a constellation or Sign. From the table observe the declination of the sun on that day. Then proceed, as described in Chapter IX, to plot the line of declination for that day on the dial plate; and place the symbol or name of the Sign or constellation at the extremities of the line.

We know of one professor of astronomy in a large college,

who uses a sundial with its lines of declination as a practical example to show the motions of the sun and earth. The students enjoy the lesson and more often than not they return many times to watch the dial, and always delight in explaining its use to their friends.

IX

HOW TO LAY OUT THE LINES OF DECLINATION

T HE succeeding pages show the construction of the lines
of declination. Rather than clutter the reader's mind with
many diagrams and lines, only those necessary for a proper
understanding of the method have been used. Although the
fundamental principle of plotting a line of declination on a
dial plate is the same for all dials, each type of dial will be
treated separately, so that the reader will have no difficulty.
In addition, the horizontal line (page 72) for each type is
shown.

Each example shows the construction of the lines representing the path of the shadow cast by the nodus (page 73)
when the sun has a declination of 0°, and when the sun
reaches its greatest northern and southern declination. The
first is often referred to as the equinoctial line, because the
shadow of the nodus falls upon it when the sun is at the
equinoxes, marking the beginning of spring and fall, when
the day and night are said to be of equal length.

The lines showing the sun's greatest northern and southern declination were called the Tropics; and on old dials
they were often labelled the Tropic of Cancer and Capricorn,

respectively. They note the longest day (beginning of summer) and the shortest day (beginning of winter) of the year. These lines may also be referred to as limiting lines, for between them all other lines of declination must fall.

It is obvious, that in all types of dials, the size, shape, and all the parts must be known, before the lines of declination can be drawn.

The Lines of Declination on the Equatorial Dial
PLATE XI

The equatorial dial may be drawn on both faces. In the example it is assumed that only the upper or north face is to be used.

Since the plane of the equatorial dial lies in the plane of the celestial equator, it is evident that all the lines of declination cannot be placed upon it. When the sun has a declination of 0° the shadow of the nodus will not fall upon the dial; and when the sun is south of the equator no shadow will be cast on the upper or north face.

If the location of the shadow of the nodus is marked when it reaches each hour line throughout any particular day and a line drawn through those points, a portion of a circle would result, with the foot of the perpendicular style as its center. For this reason it is much easier to draw the lines on an equatorial dial, than on other types.

The Construction

In Figure 1—AB represents the plane of the horizon; CD the dial plate; N the nodus; MN the height of the perpendicular style; and M the foot of the perpendicular style.

LINES OF DECLINATION ON THE EQUATORIAL DIAL
LAT. 45° N.

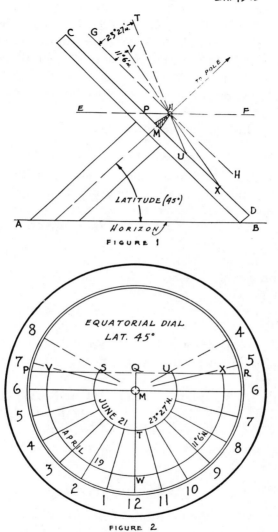

FIGURE 1

EQUATORIAL DIAL
LAT. 45°

FIGURE 2

Plate XI.

For the Horizontal Line:

Through N (Fig. 1) draw EF parallel to AB, intersecting CD at P (angle PNM is equal to the latitude of the place).

Take the distance MP and lay it off from the foot of the perpendicular style M (Fig. 2) to Q, on the 12 o'clock line. Through Q draw the line PR perpendicular to the 12 o'clock line.

Then, PR will be the horizontal line for this dial.

For the Lines of Declination:

Since the equinoxes cannot be shown on the equatorial dial, the line of declination for April 19 will be substituted.

From the table the greatest northern declination is found to be $23°27'$, on June 21; and $11°6'N$ on April 19.

Then, in Figure 1, draw the line GH parallel to CD. This represents the plane of the celestial equator.

With a protractor lay off the angle $GNV = 11°6'N$; and the angle $GNT = 23°27'N$.

Produce the lines VN and TN until they cut the dial plate, as at X and U.

Take the distances MU and MX and lay them off from the foot of the perpendicular style M (Fig. 2), to T and W, respectively.

With M as a center and radii MT and MW, describe the arcs STU and VWX, respectively. Thus will the arcs STU and VWX be the desired lines of declination.

For all other lines of declination repeat the work precisely as shown.

Sunrise and Sunset

By means of the horizontal line and the lines of declination it is possible to tell at what time the sun rises or sets.

For example, in Figure 2, the arc STU represents the path of the shadow cast by the nodus on June 21. This arc cuts the horizontal line at U. The point U lies between the hours of 4 and 5 in the morning.

Estimate the distance from the point where the hour line of 4 crosses the horizontal line, to the point at U. This will be found to be about 10 minutes. Therefore, the sun will rise at 4:10 a.m. (Apparent Time) on June 21, in latitude 45°N. An almanac computed for 45° north latitude shows the time of sunrise to be 4:12 a.m., (Apparent Time) which compares favorably with the dial reading.

Lines of Declination on the South Vertical Dial

PLATE XII

The construction of the lines of declination on the south vertical dial is typical of those dials whose planes lie oblique to the axis of the earth.

The Construction

The example shows the construction of the lines of declination for a south vertical dial in 50°N latitude.

Figure 1 shows the completed dial; the line C'12 is the substyle line. The foot of the perpendicular style is marked at F'. The horizontal line, on this dial, passes through the foot of the perpendicular style, at right angles to the substyle line.

On any dial whose plane lies oblique to the axis of the earth, the equinoctial line is a straight line, and perpendicular to the substyle line.

To find the position of the equinoctial line and other lines of declination, another diagram is used, such as that shown in Figure 2. (Early dialists called this figure a Trigon, and by it many problems of the sphere were solved.)

In Figure 2, the horizontal line PC represents the style; CF the substyle; P the nodus; PF and F, the height and foot of the perpendicular style, respectively.

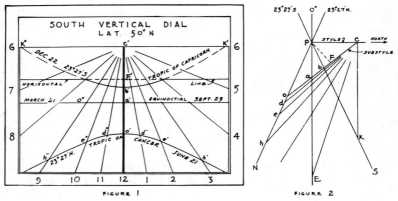

FIGURE I   FIGURE 2

Plate XII.

At the point P, draw PE perpendicular to PC. This line represents the equinoctial.

Draw PN and PS for the greatest northern and southern declination of the sun. Angles NPE and EPS are equal to $23°27'$.

Produce CF, cutting PS at b, PE at a, and PN at o.

In Figure 1, C′ corresponds to the point C, Figure 2. Lay off the distance Ca, from C′ to a′, on the substyle line C′12. The equinoctial line will pass through the point a′, perpendicular to the substyle.

With C (Fig. 2) as a center, describe arcs cutting the line PE, whose radii are equal to the distances from C′ (Fig. 1) to the points where the equinoctial line cuts the various hour

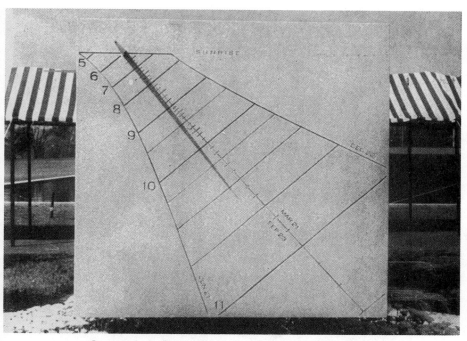

Connecticut General Life Insurance Co. Cube Dial.
The East Face is above, the South Face below.

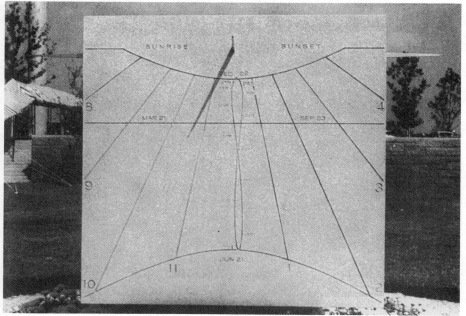

A Portable Horizontal Dial.

lines. Through the points thus found on PE, draw lines from C cutting PS and terminating in PN at o, d, e, etc. These lines represent the hour lines on the dial plate (Co = 12; Cd = 1 and 11; Ce = 2 and 10; etc.).

If the distances Co, Cd, Ce,......(Fig. 2) are plotted on the corresponding hour lines (Fig. 1) from C' to o', d', e'...., and to d'', e'',....., a curved line drawn through those points will represent the path of the shadow of the nodus when the sun has a north declination of 23°27'.

Similarly, the line K''b'K' (Fig. 1), representing the path of the shadow cast by the nodus when the sun has a declination of 23°27'S., may also be plotted by taking off the distances from C (Fig. 2) to the various points where the lines Co, Cd, Ce.... cross the line PS.

A line drawn from C (Fig. 2) perpendicular to PC and cutting the line PS at K, will give the location of the point where that line of declination intersects the 6 o'clock line, as at K' and K'' (Fig. 1). It is, however, not necessary to extend this line beyond the horizontal line.

All other lines of declination will fall between the lines K''b'K' and h''o'h' (Fig. 1); and they may be plotted as shown above by inserting the desired lines in Figure 2, making angles with PE equal to the declination, on either side of that line, as the declination is north or south.

Lines of Declination on a Horizontal Dial

PLATE XIII

The method of plotting the lines of declination on the horizontal dial is exactly the same as that used for the south vertical dial Plate XII.

Plate XIII shows the position of the lines on a horizontal dial in 40°N latitude. It is evident that the horizontal line cannot be placed on this dial, since its plane lies parallel to the plane of the horizon.

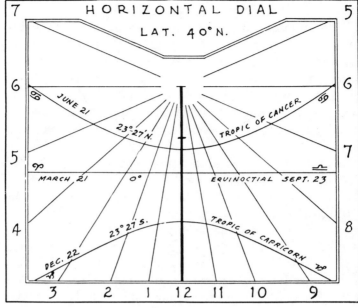

Plate XIII.

Note that the Tropic of Cancer, which is the line nearest the center of the horizontal dial, is farthest from the center of the south vertical dial. It is much easier to visualize the position of these lines if they are labelled (north or south declination), and the direction of the celestial pole properly indicated as shown in Figure 2, Plate XII.

LINES OF DECLINATION ON DIRECT NORTH AND SOUTH RECLINING DIALS

The lines of declination on the direct north and south reclining dials are constructed in exactly the same manner as those for the south vertical dial, Plate XII.

It must be remembered that the height of the style is always its elevation above the dial plate.

The location of the horizontal line is derived in the same manner as for the equatorial dial, Plate XI, and it is perpendicular to the substyle line.

LINES OF DECLINATION ON DECLINING DIALS

PLATE XIV

There are four declining dials. Of these the south dial declining east is typical. The example shows the lines of declination as they appear on a south vertical dial declining 20° east, in latitude 40°N.

The Construction

On this dial, AF is the substyle line; AB the meridian or 12 o'clock line; AN is the style; angle NAF is the height of the style; N is the nodus; NF and F the height and foot of the perpendicular style, respectively.

In order to construct the lines of declination, first, lay out the hour lines for a horizontal dial, whose meridian or 12 o'clock line is the substyle line AF; and whose height of style is equal to the angle NAF.

Then, with all parts of the gnomon for the horizontal dial equal to those of the declining dial, proceed to lay out the lines of declination as described on page 143. Thus will the lines for the declining dial be properly constructed.

The horizontal line on declining dials will pass through the point where the equinoctial line intersects the 6 o'clock line, as at K. From K, it is drawn at right angles to the 12

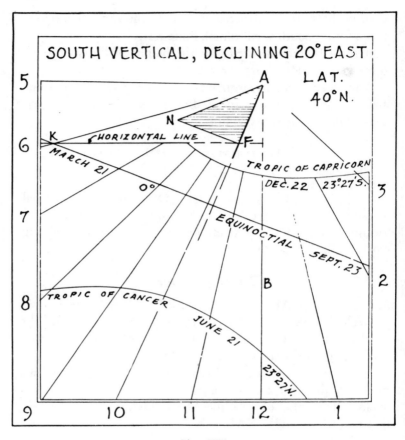

Plate XIV.

o'clock line AB. It also passes through the foot of the perpendicular style, F.

LINES OF DECLINATION ON THE NORTH VERTICAL DIAL

The north vertical dial, owing to the short period that the

sun shines upon it, is generally used only on a pillar dial. During its period of usefulness the sun has but a small elevation above the horizon. It is, therefore, not necessary to draw the lines of declination on it.

However, certain lines (depending upon the latitude of the place) may be drawn upon the dial plate if the work is done in the same manner as for the south vertical dial Plate XII.

Lines of Declination on the Direct East and West Reclining Dials

PLATE XV

The method of constructing the lines of declination on the east and west reclining dials is the same as that for the

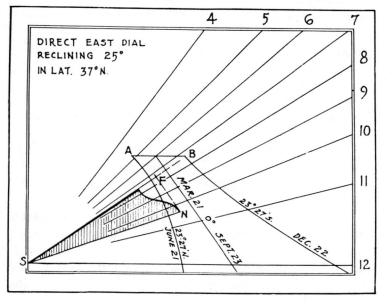

Plate XV.

declining dials, described on page 145, where a horizontal dial is inscribed about the substyle line.

The height of the style for the reclining dial is used as the height of the style for the auxiliary horizontal dial. Then, the lines of declination constructed for the horizontal dial will be the lines for the reclining dial.

The appearance of these lines on a direct east dial reclining 25°, in latitude 37°N, is shown. The line SF is the substyle; F the foot of the perpendicular style; and N is the nodus.

The horizontal line, AB, is drawn through the point where the equinoctial line cuts the 6 o'clock line and parallel to the 12 o'clock line, which on this dial is parallel to the plane of the horizon.

The horizontal line need not extend beyond the lines noting the greatest northern and southern declination of the sun.

LINES OF DECLINATION ON EAST-WEST VERTICAL DIALS

PLATE XVI

The planes of these dials lie parallel to the axis of the earth. The most satisfactory gnomon will be one shaped like a pin, the point or apex of which will serve as both style and nodus.

The method for constructing the lines is the same for both dials. The construction of the lines, described in detail, need only be shown for one of them, for example, the east vertical dial.

Figure 1 shows a direct east vertical dial computed for latitude 40°N, with the lines of declination properly drawn upon it. The most suitable gnomon is also shown. The style and nodus are coincident at N; the height of the style is equal to the height of the perpendicular style, FN; the foot of the perpendicular style intersects the 6 o'clock line at F.

Analemmatic Dial by Charles Bloud, France, about 1675.

The Construction

The equinoctial line, EF (Fig. 1), is drawn through the foot of the perpendicular style at right angles to the 6 o'clock line.

The horizontal line AB is drawn through the point where the equinoctial line cuts the 6 o'clock line and through the foot of the perpendicular style, which on this dial are coincident, at F. The line AB makes an angle with the hour lines equal to the latitude of the place (in this case 40°).

In Figure 2, the line F'E' represents the equinoctial line;

LINES OF DECLINATION ON A DIRECT EAST VERTICAL DIAL
LAT. 40° N.

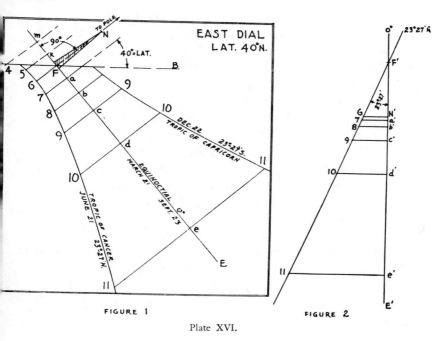

FIGURE 1

FIGURE 2

Plate XVI.

through F' draw F'11 making an angle with F'E' equal to the greatest northern declination of the sun (23°27').

Lay off from F' the distance F'N' equal to the height of the style.

Take the distances from N (Fig. 1) to a, b, c, d,.... (the points where the equinoctial cuts the various hour lines), and lay these distances off on F'E' (Fig. 2) from F' to a', b', c', d',....

Through the points N', a', b',... (Fig. 2) draw lines perpendicular to F'E', cutting the line F'11 at 6, 7, 8, 9, 10, 11, (the figures represent the corresponding hour lines in Fig. 1).

Lay off the distances N'6, a'7, b'8... (Fig. 2) on Figure 1 so that F6 = N'6, a7 = a'7, b8 = b'8,.... Then, a line drawn through the points 6, 7, 8, 9, 10, 11, (Fig. 1) will show the path of the shadow cast by the nodus when the sun reaches its greatest northern declination.

To find the points, on the hours before 6, through which the line of declination is to be drawn, make k5 and m4 (Fig. 1) equal to a'7 and b'8 (Fig. 2), respectively.

In the same manner all other lines of declination can be plotted on the dial plate.

If Figure 1 is looked at from the back, a west dial for the same latitude will be seen, with its lines of declination. The morning hours will become the afternoon hours, and the horizontal line will show the time of sunset.

Sunrise and Sunset

By combining the east and west dials, the time of sunrise and sunset may be estimated, throughout the year. The accuracy depends somewhat upon the size of the dial, and more so upon the careful laying out of the lines of declination and the horizontal line.

In Figure 1, the line of declination for December 22 cuts the horizontal line between 7 and 8 a.m., and a little before 7:30 a.m., Apparent Time. Thus, on December 22nd the sun will rise shortly before 7:30 a.m.

According to an almanac computed for latitude 40°N, the sun will rise at 7:20 a.m., Apparent Time, on December 22nd.

LINES OF DECLINATION ON THE POLAR DIAL

PLATE XVII

The appearance of the lines of declination on a polar dial is shown on Plate XVII.

The construction of the lines is the same as for the east dial.

The horizontal line for this dial has not been shown, owing to the very short period of time over which it could be used. It may be inscribed in the same manner as for the equatorial dial, Plate XI.

LINES OF DECLINATION ON THE ARMILLARY

If reference is made to the plate on page 126 the method of inscribing the lines of declination on the armillary will be apparent.

The sphere represents the plan of the heavens with the earth at its center. Therefore the nodus should be placed at the exact center of the sphere. The lines of declination will be straight lines, parallel to the plane of the equator; and inscribed on the inner surface of the circle representing the equator.

The equatorial band should be wide enough to accommodate the lines noting the greatest northern and southern declination of the sun.

The horizontal line may be determined in the same manner as for the equatorial dial, see page 140.

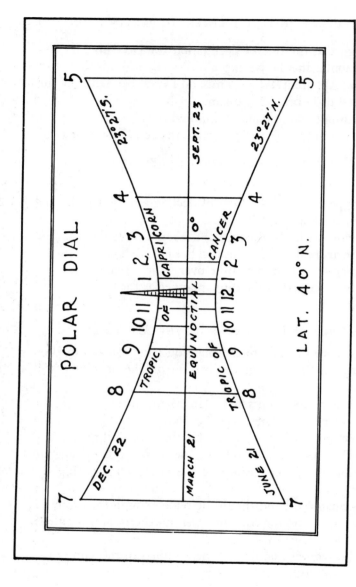

Plate XVII.

The Observatory at Jaipur, India.
After Restoration (above)
Before Restoration (below).

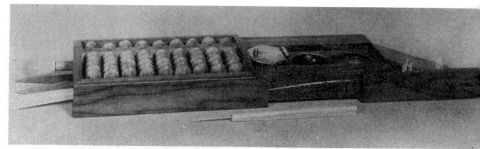

A Desirable Instrument (above)
Japanese Noon Mark Dial Designed for use at Sea (below)
From Ernst Collection

Plotting Lines of Declination by the Altitude Method
PLATE XVIII

There is another method of plotting the lines of declination, which employs tables showing the altitude of the sun. This will no doubt appeal to those who have access to such tables.

The most satisfactory tables are those published by the U.S. Hydrographic Office, designated as No. 201 and No. 203. These tables show the altitude and azimuth of celestial bodies for stated values of declination and latitude, and they will be found very useful when constructing a sundial.

Note: Publication No. 201 is now out of print and consequently difficult to obtain; but No. 203 is the current publication and contains the same tables.

The application of this method to the various types of dials is the same as that shown in the following example, where the path of the shadow cast by the nodus is plotted on a horizontal dial in latitude 40°, when the sun has a declination of 20°N.

From the tables mentioned above, take out the values for the altitude (angular distance above the horizon) of the sun, for each hour of the day, thus:

THE SUN'S DECLINATION FOR APPARENT NOON

Hour (*Apparent Time*)	Altitude of Sun
12 noon	70°
1 p.m. and 11 a.m.	66° 14′
2 " " 10 "	57° 29′
3 " " 9 "	46° 48′
4 " " 8 "	35° 26′
5 " " 7 "	23° 58′
6 " " 6 "	12° 42′

(The values shown above, were obtained from page 412, U. S. Hydrographic Office Publication No. 201.)

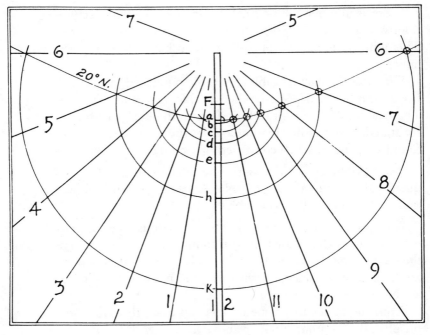

FIGURE 1

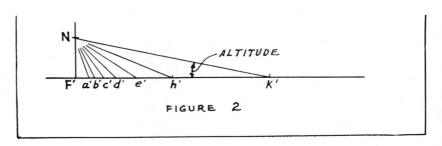

FIGURE 2

Plate XVIII.

In Figure 2, the line F′k′ represents the substyle line on the dial. The foot of the perpendicular style is noted at F′, and the nodus at N. NF is the height of the perpendicular style.

Draw lines from N to the line F′k′, making angles with F′k′ equal to the altitudes shown in the table above. Thus, angle F′a′N = 70°; angle F′b′N = 66° 14′; . . . .

Then, in Figure 1, lay off from F (the foot of the perpendicular style) the distances Fa, Fb, Fc, . . . equal to F′a′, F′b′, F′c′, . . (Fig. 2) respectively.

With center at F (Fig. 1) and radii Fb, Fc, Fd, . . . describe arcs cutting the corresponding hour lines (radius Fb cuts the hour lines 1 and 11; Fc, 2 and 10; and so on).

Through the points thus found on the hour lines, draw a curved line, which will be the desired line of declination.

All other lines of declination may be plotted in the same manner.

X

PORTABLE SUNDIALS

THERE was no more glamorous period in timekeeping than that from the 15th to 19th centuries when the craftsmen of the period expressed their art in portable sundials. These were the pocket watches of yesteryear. They rarely exceeded 4 inches in their largest dimension and most of them could be put in one's pocket.

Portable dials were fashioned out of every conceivable material from paper and wood to ivory, silver, and gold; and they were constructed in every imaginable form—armillaries, rings, cubes, crosses, books, tablets, disks, ships, cylinders, and guns. The dials of Schissler, Habermel, Tröschel, Engelbrecht, Tucher, Reinman, Bloud, Butterfield, and many others are beautiful specimens of the craft, exhibiting perfection in delineation and ingenuity in design. Even the simplest of these dials has a rare charm and beauty. Not only are they fine examples of artistry and engraving, but they are remarkable for their accuracy.

Over the years many people have found making portable dials to be such a fascinating hobby that we have included construction details for some.

The ordinary horizontal dial was a common portable type. One to five hour bands were laid out for intervals of three to four degrees of latitude to extend the geographical range over which a traveler could use the dial. A folding gnomon was adjustable to correspond to each latitude band. Sometimes a horizontal and a vertical dial were constructed and fitted with a hinge so that the two tablets could be folded together. Such devices were called "book dials" or "diptych dials".

Unlike the horizontal and other dials (stationary or portable) described in preceding chapters, there are many interesting dials that tell time by measuring the altitude of the sun (angular height above the horizon). Typical of this type are the so-called "pillar dials" and "card dials", shown facing pages 81 and 161. Other altitude dials will be found in the form of disks, quadrants, and so forth.

Still other interesting dials tell time by measuring the azimuth of the sun (angular distance, in time, measured on the horizon between the south point and the foot of the perpendicular from the sun). Some of these are designed to be used with other types to obviate the necessity of a compass to set them.

You must remember that many of these dials were devised long before mechanical clocks were known; and that ancient peoples placed great importance on the position of the sun in the signs of the zodiac. For that reason we find seven lines on many of these dials, which represent the date of the sun's entrance into the "signs". The dates are determined from the sun's longitude, measured in degrees eastward from the vernal equinox along the ecliptic, not along the equator. The sun enters a new sign about every 30 days. Although the signs no longer coincide with their respective constellations, we continue to use the names of the old zodiacal signs.

The signs being of equal length, the seven lines on the dials are parallel and equally spaced. The following table gives the name of each sign, date of the sun's entrance, and its declination on that date. The arrangement of the table shows why only seven lines are required.

Sign	Ent. Date	Dec.	Ave. Dec.	Dec.	Ent. Date	Sign
Cap	Dec. 22	$-23°\ 27'$				
Aqr	Jan. 20	$-20°\ 16'$	$-20°\ 08'$	$-19°\ 59'$	Nov. 22	Sgr
Psc	Feb. 19	$-11°\ 31'$	$-11°\ 20'$	$-11°\ 09'$	Oct. 23	Sco
Ari	Mar. 21	$0°\ 00'$	$0°\ 00'$	$0°\ 00'$	Sep. 23	Lib
Tau	Apr. 20	$+11°\ 17'$	$+11°\ 29'$	$+11°\ 41'$	Aug. 23	Vir
Gem	May 21	$+20°\ 02'$	$+20°\ 08'$	$+20°\ 13'$	July 23	Leo
				$+23°\ 27'$	June 21	Cnc

For the full constellation names and their symbols, see page 135.

Note that except for the winter and summer solstices the sun has approximately the same declination at its entrance into two signs, such as Aquarius in January and Sagittarius in November. The arrangement shows a symmetry on either side of the solstices. The calendar distance between the beginnings of the signs varies only by a day. For example: the number of days from November 22 to December 22 is 30; from December 22 to January 20 is 29. This difference is so small that it is disregarded; therefore one line is used for the two dates, or signs.

Also, as you can see, the difference in declination is so small that we can take the average, as given in the middle column. Therefore, we can construct a short table of the sun's azimuth and altitude for each hour of the day at the date of the sun's entrance into the signs. Note too how nearly symmetrical the average declination is about the equinox. Because we are dealing with very small dials that can be put in your pocket, we can round off the average declination to the nearest 30'.

In order to make it easier for you to get started constructing some of these dials, we refer you to certain publications which may be obtained from the Superintendent of Documents, U.S. Government Printing Office, Washington, D.C. 20402.

H. O. Publication No. 214: *Tables of Computed Altitude and Azimuth*. This is a set of nine volumes, each volume giving the altitude and azimuth for 10° of latitude, and each degree (4^m) of hour angle. Vol. V covers latitudes +40° to +49°, inclusive; Vol. IV, +30° to +39°, inclusive. This publication replaces an older but very useful single volume H. O. No. 201: *Simultaneous Altitudes and Azimuths of Celestial Bodies,* which may be found in second-hand bookstores.

H. O. Publication No. 260: *Azimuths of the Sun.* This volume contains the azimuths of the sun at intervals of 10 minutes of time from sunrise to sunset for each degree of latitude from 0° to +70°, inclusive, and for each degree of declination from 0° to ±23°, inclusive.

If you do not have access to these books, formulas for computing azimuth and altitude are in the appendix.

In all cases the constructions shown here are for latitude 40°N and concern those elements that cannot be changed. The overall design of the dial is left to your own initiative.

The Ring Dial

PLATE XIX

The basic construction is the projection of the altitude of

the sun at its entrance into the signs of the zodiac. The indicator is light, which passes through a minute hole in the ring to the opposite inner surface whereon the hour lines are inscribed. These dials are quite small, seldom more than an inch in diameter, from 3/8 to 3/4 of an inch in width, and from 1/32 to 1/16 of an inch thick. To tell the time, the ring is suspended and the hole turned toward the sun until a ray of light falls on the date line, in which position the spot of light will indicate the hour.

Construction (Fig. a)

First determine the size of the ring—its diameter, thickness, and width. Draw the ring as in (Fig. a). Draw the

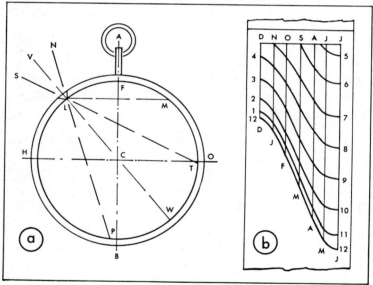

Plate XIX.

Armillary on the Campus at Phillips Academy, Andover, Mass.

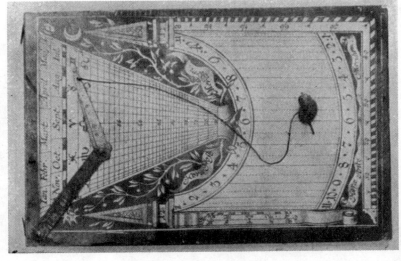

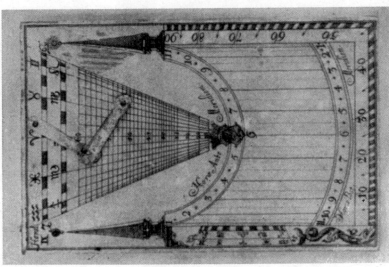

Two Examples of Capuchin Portable Dials. Both are about Three Inches Wide.

axes AB and HO perpendicular to each other, through the center C.

Let F on AB be the point of suspension, then HO represents the horizon.

From the center C draw a line VC making an angle with HO equal to the co-latitude (90° minus latitude) of the place. Where the line VC intersects the inner surface of the ring at L, drill a small hole and bevel the outside surface, as shown. Do not make the hole too large—not much larger than a pinhole.

Extend VC to the inner surface at W, thereby marking the position of the spot of light at noon, at the equinoxes.

Draw LM parallel to HO. The point M represents sunrise and sunset.

From L draw lines NLP and SLT, each making an angle with LCW equal to 23 1/2°. The points P and T are the positions of the spot of light at noon at the summer and winter solstices, respectively.

The angles MLT, MLW, and MLP are equal to the altitude of the sun at noon on the respective dates. The remaining hours of the day are located by laying off lines from L that make angles with LM equal to the altitude of the sun for each hour. Repeat the process for each of the dates of the sun's entrance into the signs.

Once the hour points have been determined for each date, they must be transferred to the dial plate on the inner surface of the ring. (Fig. b) shows the appearance of the dial plate if the ring were cut and flattened.

An easy way to make your first dial is to use a thin cardboard tube, stiff enough to hold its shape. Many household accessories such as paper towels or wrapping foil will provide a satisfactory tube of about the right diameter. Then cut a piece of paper the width of your ring and a little longer than the arc PWTM.

Draw seven lines equally spaced and plot the hour points. Then paste the paper dial plate on the inner surface of the tube, in the correct position.

THE CYLINDER OR PILLAR DIAL

PLATES XX AND XXI

Cylinder dials have been used for centuries. They were sometimes referred to as "poke dials" because the shepherds carried them in their poke. In some parts of the world today these dials are still employed, such as in the Pyrenees. The shepherd's dial was made of boxwood about 1/2 to 3/4 of an inch in diameter and about 3 to 4 inches long. The gnomon is usually constructed so that it may be folded down into the cap for storage. The dial tells time by setting the gnomon over vertical lines representing the days of the year. The cylinder is then suspended and turned until the shadow of the gnomon is vertical or parallel with the date lines. The shadow of the nodus or point of the gnomon among the hour lines indicates the time.

Construction (Plate XX)

The construction of a cylinder dial is both easy and interesting. The diameter of the cylinder may be anything you wish. Let's start out with a 2-inch cylinder. The style and nodus are coincident in this dial. The length of the cylinder is governed by the distance of the nodus from the surface of the cylinder. Let's assume this to be 1 inch. The gnomon must be constructed so that it can rotate about the cylinder.

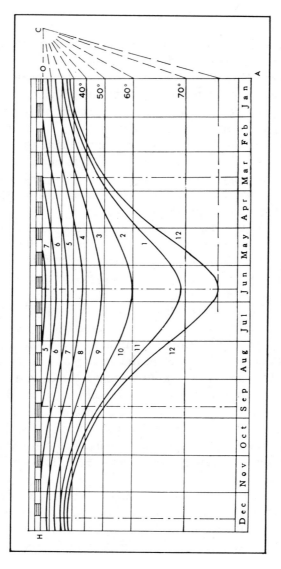

Plate XX.

This can be accomplished by screwing a thin cap to the top of the cylinder and attaching the gnomon to it.

We need to know the circumference of the cylinder, which is found by the formula $2\pi R$, or by multiplying the diameter by 3.1416. A 2-inch cylinder will have a circumference almost exactly 6 1/4 inches. Our diagrams were drawn on the basis of these figures, and we have assumed that the hours and date lines will occupy the entire circumference, although this is not necessary. You can lay them out using only 5 inches or 5 1/2 inches. The length of the cylinder will be determined from the construction of the hour lines.

First, draw HO for the horizon line. The distance between H and O should be equal to the circumference of the cylinder, or whatever distance you decide. Extend HO to C, the distance OC equal to the length of the gnomon, and the point C will be the nodus.

The distance HO will be equal to 365 days, which must be divided into months. The segmented bar above HO marks the 1st, 10th, and 20th of each month, and the long lines perpendicular to HO represent the 1st day of each month. Also locate the dates of the equinoxes and solstices, and from them drop perpendiculars (long broken lines) from HO.

Lay a protractor on HOC, centered at C, and mark off angles every 10°, or better still 5°. Then draw lines to C that intersect OA, the 1st of January. From the intersection points on OA draw lines parallel to HO. These will be lines of altitude.

Next prepare a table of declinations, in whole degrees, for the first of each month. For example: January 1 is —23° 04′, January 2 is —22° 59′; therefore you can adopt —23° for January 1. However, you may need to draw an

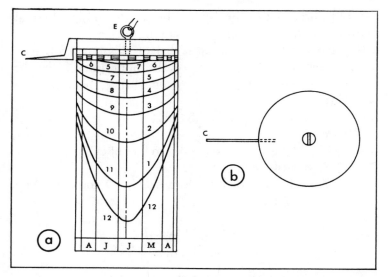

Plate XXI.

extra date line or two, such as for September 5 when the declination is $+7°\ 05'$, which rounds to $7°$.

Now you can obtain the altitudes for each hour of the day for each declination in your table. Then plot the altitudes on the grid formed by altitude and date lines. Connect the points with smooth curves and you will have constructed the hour lines appropriate for this dial. All that is left to do is to label the months and the hour lines. Then cut out the completed dial plate and paste it on the cylinder.

The appearance of a portion of the dial is shown on Plate XXI (Fig. a). Note that the style or nodus C must be perpendicular to the surface of the cylinder and on the horizon line. (Fig. b) shows the top of the dial. A ring, E (Fig. a), should be attached to the top, in the center, so that the dial may be suspended in a vertical position.

If you want to make a large dial of this type, you may wish to compute the distance of the hour lines from the horizon line HO. This may be done using the following formula:

$$X = HS \tan A$$

Where

HS = height of style (nodus to face of cylinder)
A = altitude of sun
X = distance of hour line from HO corresponding to date

DIAL ON A QUADRANT

PLATE XXII

The early dials of this type were made on thin plaques not more than a quarter of an inch thick, and some were fashioned out of thin metal, hence the name "card dials". They can also be made on a piece of stiff cardboard that is cut in the shape of a quadrant, or the quadrant can be delineated on a square or oblong.

These dials may be constructed either for use at a specific place or for any locality. When shade or a ray of light is properly oriented along a specific line, time is told by means of an auxiliary indicator in the form of a bead or knot in a piece of thread. The dials are held vertically, and there is no need to know the location of the meridian.

Construction (Plate XXII)

The "quadrant dial" is the simplest and easiest to make.

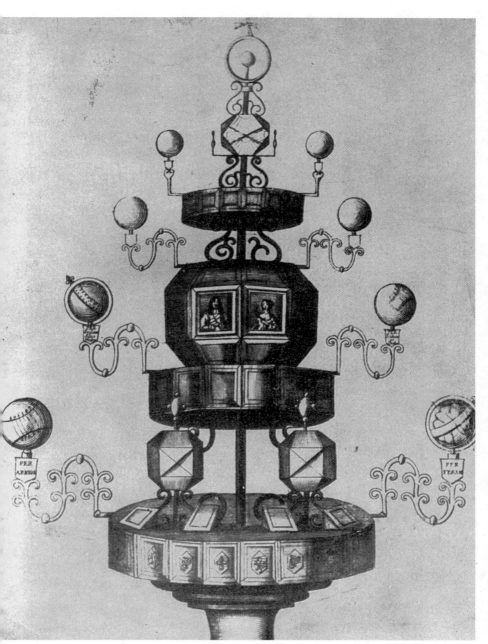

The Whitehall Dial.

Floral Sundial, San Francisco, Calif.

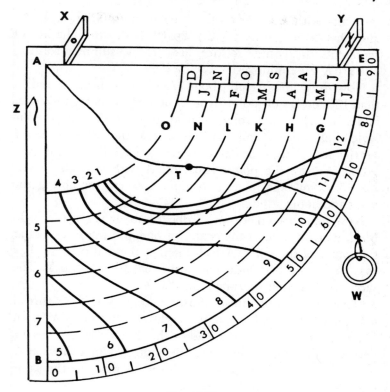

Plate XXII.

Since its construction is based on the sun's entrance into the signs, seven equally spaced lines are used. Of course you can use date lines if you wish; but, in any case, the winter and summer solstice will govern the length of the date scale.

First draw two lines AB and AE perpendicular to each other. With A as a center, draw arc BE of any convenient radius, say 3 or 4 inches.

Again, with A as the center draw six arcs, G, H, K, L, N, and O, 1/4 or 3/8 of an inch apart, as shown. The arc O represents the winter solstice, arc BE the summer solstice, and arc K the equinoxes.

Now draw another arc about 1/4 of an inch larger radius than BE and lay off points from B at 5 or 10 degree intervals, and mark them as shown from 0 to 90. This will be your scale of altitudes. Using this scale with a ruler set at A lay off on the arcs of signs or dates the respective altitudes for each hour. Connect the points thus found with smooth curved lines and your dial plate is complete, except for noting the dates and hours.

Now make two pieces of cardboard about 1/4 x 1/2 inch. Make a small hole in the center of one and mark a cross at the center of the other. Attach them to the dial plate so they may be erected perpendicular to the dial as at X and Y. The piece with the hole must be located at X.

Obtain a piece of thread and tie a knot as at T, or thread a small bead. Attach a small weight or plummet, W, such as a ring or washer, at one end of the thread. With a needle pass the other end of the thread through A. Cut a slit as at Z, which can be used to hold the thread in place in the same manner as it is held on its spool.

If a bead is used, it should be tight enough to stay in place once it is set. If a knot is used, the thread must be long enough to allow adjustment of the knot over the entire length of the date scale.

To use the dial, first set the bead or knot to the proper date or sign. Then hold the dial plate in a vertical position and point the hole at X toward the sun. Be sure the thread moves freely across the dial plate. When the sun's rays passing through the hole are centered on the cross at Y the knot or bead will indicate the time; and the position of the thread as it crosses the altitude scale will give the altitude of the sun.

If you do not want to use the sights shown at X and Y, devise a way of attaching a needle so that it will be perpendicular to the dial plate on the line AE about at X. Then with the dial in a vertical position and in the plane of a line between you and the sun, allow the shadow of the needle to fall along the line AE, in which position the bead or knot will indicate the time.

Now let's try another kind, often referred to as a "Capuchin dial," because of its resemblance to a monk's hood.

THE CAPUCHIN DIAL

PLATE XXIII

This dial is constructed in a similar manner to the quadrant dial, but its appearance is quite different. You can lay it out on a stiff piece of cardboard, or you can make it first on paper that is later pasted to cardboard or to a thin piece of wood. Cigar box wood is good material for making these dials. Sights or an alidade can be used, as for the quadrant dial, or as shown here. A weighted thread, fitted with a bead or knot, is passed through a slit in the date scale, to serve as a secondary indicator.

Construction (Plate XXIII)

Let abcd represent the card.

Draw AB parallel to the top of the card. At C on AB draw CE perpendicular to AB.

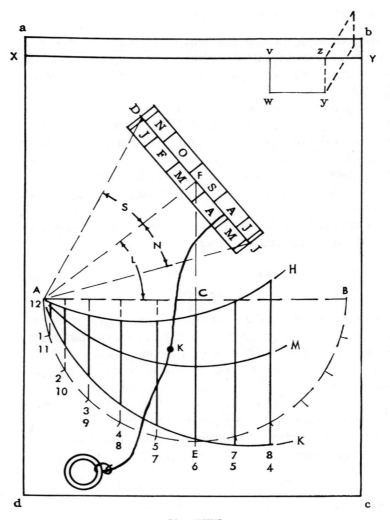

Plate XXIII.

With C as a center and any convenient radius, describe the semicircle AEB and divide it into 12 equal parts or segments of 15°. From the points thus found draw lines perpendicular to AB.

The point A marks 12 o'clock, CE the 6 o'clock line. By further subdividing the semicircle, the half and quarter hours may be drawn.

At A draw a line making an angle L with AB equal to the latitude of the place and intersecting the extension of CE at F. Through F draw a line perpendicular to AF. This will make the scale of dates.

Draw two lines from A, each making a 23 1/2° angle (N and S) with AF, which equals the declination of the sun at the solstices. The intersection of these lines with the date scale at J and D, respectively, will determine the limits of the date scale; angle N applies to north declinations, angle S to south ones. The point F is the position on the scale of the equinoxes.

The old dials of this type made use of the seven lines or divisions of the scale to represent the sun's entrance into the signs; but we shall show only the lines for the equinoxes and solstices and the division of the scale into months.

With D on the date scale as a center and radius DA, describe an arc AH, which will represent the path of the indicator (bead or knot) at the winter solstice. With J as a center and radius JA draw arc AK, the path of the indicator at the summer solstice. The hour lines are required only between these two arcs. With F as a center and radius FA, describe the arc AM, the path of the indicator at the equinoxes. You can leave arc AM out, if you want to.

Now draw XY parallel to AB, and your dial is complete,

except for noting the hours, dates and/or signs, and constructing the gnomon.

The small rectangle vwyz can form the gnomon by cutting along the three sides vz, vw, and wy. Score the side yz to make a hinge. The gnomon can then be pushed up out of the card into a vertical position, and when not in use it can be set back in its pocket in the card. Cut a slot along DJ, in the date scale, to receive the thread.

To use the dial, set the thread in the slot DJ opposite the day of the month or sign. Stretch thread across the point A or 12 o'clock, and slide the bead, K, or pull the thread until the bead or knot is on the Point A. Set the gnomon in position, perpendicular to the dial plate.

Now hold the card vertically, pointing the gnomon toward the sun, tipping it in the vertical plane until the side vz of the gnomon casts a shadow along the line XY. Then the position of the bead or knot among the hours will tell the time.

Universal Capuchin Dial

PLATE XXIV

This too is called a Capuchin dial, though it is designed for any latitude. Therefore, for rigidity we suggest the dial plate be pasted onto a piece of hardwood about a quarter of an inch thick. A weighted thread is used, as in the previous dials, but in this case it must swing from the latitude line corresponding to the geographical location of the dial. A small pin, such as a map pin, can be used to hang the thread.

Construction (Plate XXIV)

First determine the size of the dial you want to make,

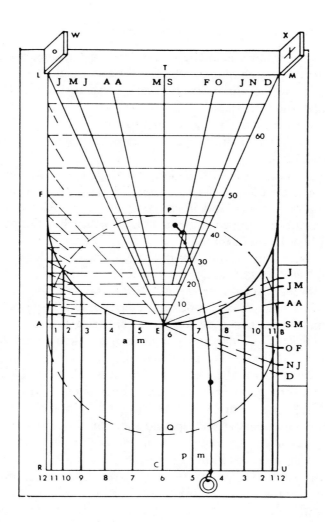

Plate XXIV.

say 4 inches wide. The construction determines the length. Allow enough room on the side for a date scale.

Draw the horizontal line AB, equal to the width of the dial, and bisect it at E.

At E draw CT perpendicular to AB.

At A and B draw lines RL and UM respectively, parallel to CT.

From E draw lines making an angle of 23 1/2° with ET, and intersecting AL at L, and UM at M. Draw LM for the top of the dial, and intersecting ET at T. Also draw lines from E making an angle of 23 1/2° with EB and intersecting BM at J, and BU at D.

With E as a center and radius EA draw a circle intersecting ET at P, and EC at Q.

With P as a center and radius PE draw a semicircle, which will be the upper limit of the hour lines. The lower limit may be found by describing an arc with E as a center and radius ED. Then allow a little space below this arc and draw RU for the bottom of the dial.

Now divide the arc AQB into 12 equal parts of 15° each, and through these points draw lines parallel to CE. These will be the hour lines, which should be marked as shown, the a.m. hours at the top, and the p.m. hours at the bottom.

Lay off on the date scale JD the declination of the sun at its entrance into the signs (see page 158). The point D represents the winter solstice, B the equinoxes, and J the summer solstice. All south declinations will lie within the angle BED, all north declinations within the angle BEJ.

Now we must plot the same declinations on LM. The point M will be the winter solstice, T the equinoxes, and L the summer solstice. Draw lines from E to the points on LM.

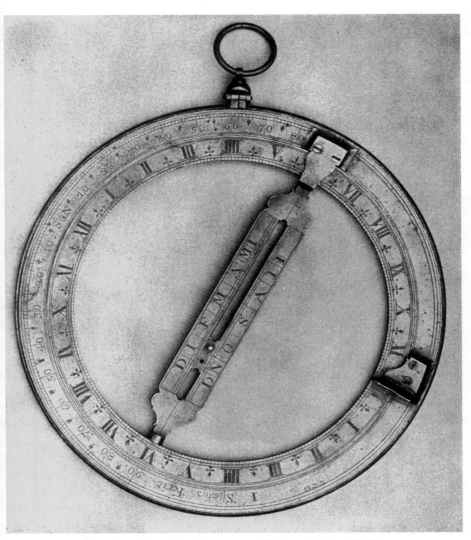

Jonathan Sisson's 18th-century Universal Ring Dial.

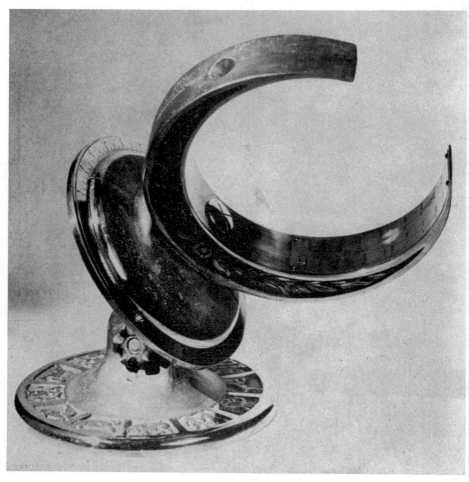

A Standard Time Dial by Edwin Griffiths.

These will represent the sun's entrance into the signs, and the reason for drawing them in this manner will become apparent. Lay a protractor on AEB centered at E, and divide the arc AP into 5° or 10° segments. Through these points draw lines from E intersecting AL. From the points thus found on AL draw lines parallel to AB, between EL and EM. These will be the lines of latitude.

The dial is now complete except for noting the hours, dates, degrees of latitude, and adding sights at W and X, described heretofore.

To use the dial, stick a pin with thread attached on the line corresponding to your latitude (in this case 42 1/2° N) and date. Stretch the thread across the date scale JD and set the bead over the proper date. Then hold the dial in a vertical position and point the sight W toward the sun. When the ray of light is centered on the cross on sight X, the bead will indicate the time.

If you want to make the dial look like the old ones, divide each of the signs into three equal parts, thereby making the lines that cross the latitude lines approximately the 21st, 1st, and 10th of each month. Divide the date scale in a similar manner.

The diagram shows the way the dial was made in the "old days". However, today it will be more meaningful if the signs are transferred into months. We have not shown all the lines that usually are inscribed on these dials in order to make the method of construction clear.

Also you will notice that when the pin is properly set to latitude and date, the thread will indicate the time of sunrise and sunset when the top of the dial LM is held in a horizontal position.

THE O-G DIAL

PLATE XXV

This dial is distinguished from the other altitude dials by the resemblance of the hour lines to the ogee moulding on a cornice. It is included here to show the many different ways the same construction was employed by early dial makers. A weighted thread fitted with a bead or knot is used.

Construction (Plate XXV)

Draw the horizontal line AB. From B draw BC perpendicular to AB. Let BC equal the length of the dial.

With B as a center and radius BC, describe arc CQ. Lay off points on CQ, from C, at 5° or 10° intervals. This will provide an altitude plotting scale.

Now pick point G about 1/3 the distance from B to C. Divide CG into 6 equal parts, which will represent the signs. The point C is the winter solstice, H the equinoxes, and G the summer solstice.

With B as a center draw concentric arcs from the division points on CG, as shown. These arcs may be labeled with months, as indicated, or with signs.

Lay off the hours by setting a ruler against B and the altitudes on CQ. Their intersection with the corresponding sign arcs will locate the hours for that date. Repeat for each arc, then draw smooth curves through the points found. Draw AE parallel to BC.

The dial is now complete except for noting the hours and dates and providing sights, or a needle for casting a shadow. The needle should be located at B. If sights are

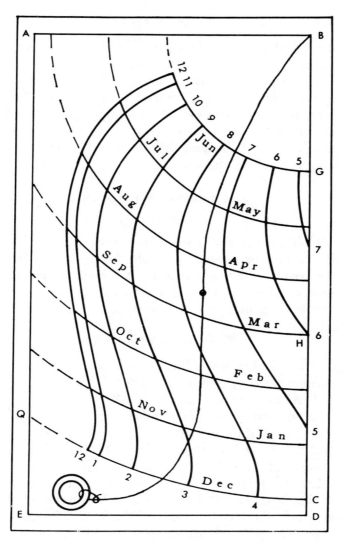

Plate XXV.

used, drill a hole in the one at B and mark a cross on the one at A.

To use this dial, insert a thread at B and adjust the bead or knot to the proper arc or date. Hold the dial vertical and turn the edge BC toward the sun, tipping it until the ray of light falls on the cross, or the shadow of the perpendicular needle or pin falls along AB. In this position the bead or knot will indicate the time.

HORIZONTAL ALTITUDE DIAL

PLATE XXVI

Unlike the previous dials, this one is used in a level position. It tells time by measuring the altitude of the sun, but it has one disadvantage—it cannot tell the hour nearest sunset or sunrise. This dial uses a perpendicular gnomon, such as a needle, the point of which is the style.

Construction (Plate XXVI)

Draw the horizontal line DC and assume point A as the position of the style, the height of which is 1 inch.

At A draw AG perpendicular to DC.

Divide both quadrants into three equal parts of 30° each, and from A, through those points, draw lines AE, AF, AH, and AP. These lines together with DAC and AG represent the dates of the sun's entrance into the signs.

Let AC represent the sun's entrance in December, the winter solstice; AP in January and November; AH in February and October; AG in March and September, the equinoxes; AF in April and August; AE in May and July; and AD in June, the summer solstice.

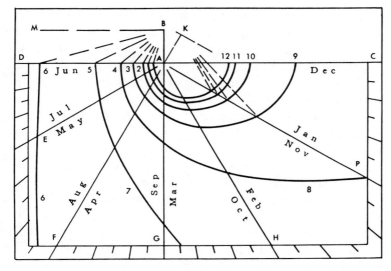

Plate XXVI.

Extend AG to B making AB equal to the height of the style, in this case 1 inch. The point B will be the style and nodus, which are coincident.

First draw the line BM parallel to DA. Lay a protractor on BM centered at B. From BM lay off the altitudes of the sun for each hour at the summer solstice. Draw lines from these points to B. Their intersection with DA will locate the hours on that date. Repeat for the winter solstice.

At A erect AK perpendicular to AP and equal to AB. From the style K draw a line parallel to AP. Again lay off the altitudes for that date, connecting them with K. Their intersection with AP will locate the hours on that date. Repeat the process on lines AH, AG, AF, and AE.

Draw smooth curves through the hour points and the dial plate will be complete except for noting the hours and dates. Affix a gnomon perpendicular to the dial plate at A.

To use the dial, lay it on a level surface and point the gnomon toward the sun so that its shadow falls along the line corresponding to the date. Then the shadow of the style, the point of the needle, will indicate the time.

In order to use the dial on dates other than the sun's entrance into the signs, the periphery of the dial can be divided into 5° or 10° segments. Count the days from the sun's entrance, then point the shadow to the number of degrees. For example, suppose you want to use the dial on April 10. Since the sun's entrance into the signs was usually considered the 21st of the month, the number of days will be 20 since March 21. Point the shadow to 20° and the dial will be set for April 10. Of course you can plot the days of the month as well as the signs.

Dials of this sort may be laid out very easily by using trigonometry. In the diagram, the distance from A to 12, A to 1, and so forth may be found from this formula:

$$X = HS \cot A$$

Where
  HS = height of the style
   A = altitude of the sun
   X = distance of hour points from the foot of the perpendicular style

HORIZONTAL AZIMUTH DIAL

PLATE XXVII

This dial is fixed in a horizontal position with its 12 o'clock

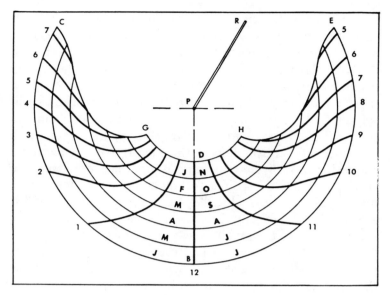

Plate XXVII.

line in the plane of the meridian. The shadow of a fixed gnomon or style is used as an indicator. The basis of its construction is the sun's entrance into the signs.

Construction (Plate XXVII)

Draw the vertical line PB.

With P as a center and radius PB describe the arc CBE, which will represent the date of the sun's entrance into the sign of Cancer, the summer solstice, June 21.

On PB mark the point D so that PD equals about 1/3 of the distance PB.

With P as a center and radius PD, describe the arc GDH,

which will represent the winter solstice, December 22.

Now divide DB into 6 equal parts and with P as a center describe concentric arcs through those points. These arcs will represent the dates of the sun's entrance into the other signs, the middle arc being the equinoxes.

Set a protractor on PB centered at P and lay off on each side of B the azimuth angle of the sun for each hour on June 21. The segments representing hours are symmetrical on either side of PB. Repeat the plotting of the azimuth angles for each of the remaining signs. Then draw smooth curves, through the points for the hour lines. More arcs may be added if desired. Label the hours as shown, and fix a thin style PR vertically to the dial plate at P.

The length of the gnomon will depend on the size of the dial. A good proportion will be the distance from P to the equinoctial arc. However, the shadow will not fall on the outer arc between 9 in the morning and 3 in the afternoon, approximately, unless the height of the gnomon is proportioned to the altitude of the sun at noon or one hour after noon on June 21. But this makes the gnomon look too high for the dial. Therefore, the shadow of a shorter gnomon usually is extended by eye to the date arc during the period when the shadow does not fall on the arcs.

To use the dial, place it on a level surface with the gnomon toward the south and the 12 o'clock line in the plane of the meridian. In this position the shadow of the style on the arc corresponding to the date will indicate the time. On many dials of this type the intervals between arcs can be divided into three parts, thereby making it easier to observe the position of the shadow. The arcs of the sun's entrance into the signs should be solid lines, the dividing arcs dotted.

MITTENWALD, GERMANY

A Madonna is the focal point of this dial on the wall of a church. Note that, for no apparent reason, the a.m. hours of 7 and 8 o'clock are lettered upside down relative to the rest of the dial.

STAMFORD, CONNECTICUT

The effective use of contrasting colors and perspective give this wall dial a striking three-dimensional appearance.

Color may be used to differentiate between the hour lines and the subdivisions.

MAGNETIC AZIMUTH DIAL

PLATE XXVIII

The magnetic azimuth dial is an ingenious device for telling time by means of a magnetic needle, such as is found in a compass. Most portable horizontal azimuth dials and many other dials make use of a small compass set into the dial plate to orient the dial on the meridian. The magnetic azimuth dial reverses the process—the needle is the indicator. But we know that a compass needle does not point to the true north except in certain places. The amount it varies from true north is called its declination or deviation. The amount of deviation varies from place to place, from year to year. It may be east or west depending upon your location. Therefore the magnetic azimuth dial will not show the correct time; but early dial makers had a way of making it work, which we will explain after we have shown the construction.

Construction (Plate XXVIII)

The hours on the dial plate for this dial are constructed in exactly the same manner as the horizontal azimuth dial previously described and shown on Plate XXVII. However, the numbering of the hour lines must be reversed, as will become apparent.

Instead of fixing a vertical gnomon at point P, fix a pinion upon which a magnetic needle can float and swing freely, suspended close to the dial plate.

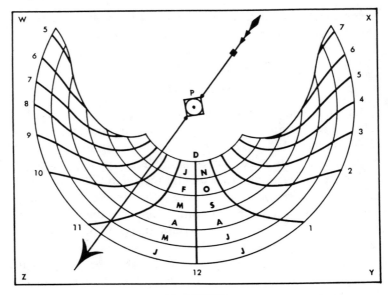

Plate XXVIII.

To use the dial, place it on a level surface so that the side WX is toward the sun. Turn the dial until the sides XY and WZ do not cast a shadow, then the time will be indicated by the position of the needle on the arc corresponding to the date.

Now, we have assumed in this construction that the magnetic needle points to the true north; but, as we pointed out above, for most places on earth it deviates from true north.

Suppose you live in a place where the magnetic declination is 15° west. The way to make this dial tell more nearly the correct time is to design the outline of the dial so that when it is noon the magnetic needle will lie along the 12 o'clock line. This is accomplished for west declination by having the side WZ make an angle with ZY equal to the complement of the deviation (90°—15°) or in this case 75°. The side XY should be made parallel to WZ making an angle with YZ equal to 90° plus the deviation, or in this case 105°. Then when the sides WZ and XY do not cast a shadow, the time indicated by the needle will be more nearly correct.

Another method used to adjust for magnetic declination was to construct the dial plate in a circle set into a square block. The dial plate was then turned in the direction of the declination by an amount equal to the deviation. This procedure results in the same relationship as cutting the sides at an angle.

In the early days of dialling, these instruments might have worked quite well; but today it is hard to get away from powerlines, steel, and other structures that affect a compass. Therefore, today's magnetic azimuth dial is more a curious and ingenious device than a practical one.

XI

VARIABLE CENTER DIALS

T HERE are two interesting dials of this form—the so-called Analemmatic Dial, and the Lambert Dial named after M. Lambert who devised it in 1777. These dials make use of a moveable gnomon, and both use the azimuth of the sun for telling time.

They may be constructed as stationary or portable dials. Both are usually combined with an ordinary horizontal dial, thereby making it possible to obtain proper orientation on the meridian without recourse to a compass, or previous knowledge of the location of the meridian. This is possible because the combination of two dials using different methods of telling time, such as hour angle and azimuth, hour angle and altitude, or azimuth and altitude, enables you to turn the instrument until the two dials record the same time; then they will lie in the meridian.

The Analemmatic Dial
Construction (Plate XXIX)

Draw the horizontal line WX in (Fig. a); and YZ perpendicular to WX, intersecting at O. YZ represents the

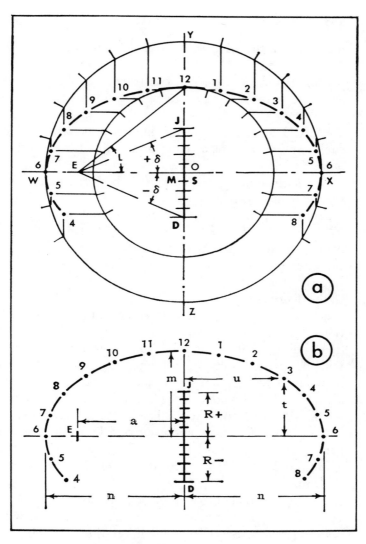

Plate XXIX.

meridian of the dial, or 12 o'clock. The line WX is the 6 o'clock line.

From any point E on WX construct a line from E making an angle with WX equal to the latitude of the place and intersecting YZ at 12.

With O as a center and radius O-12 describe the inner circle; and with O as a center and radius E-12 describe the outer circle.

Divide the two upper quadrants of both circles into six 15° segments, as shown.

From the points on the outer circle draw lines parallel to YZ, and from the points on the inner circle draw lines parallel to WX. The intersections of these two lines will be the hour points, which if connected with a smooth curved line will form an ellipse. Mark the hour points as indicated in the diagram.

Set a protractor on WX centered on E. Lay off the declinations (δ) of the sun for the days desired. Draw lines through these points to intersect YZ. Note that north declinations will be plotted between O and 12; south declinations between O and Z. The point J represents the date of the summer solstice; M and S the equinoxes; and D the winter solstice. Plot as many intermediate dates as you desire.

Now make a thin gnomon out of a knitting or common needle, constructed so that it may be moved back and forth along the date scale in a vertical position. Now your sundial is complete.

To use the dial, place it on a level surface so that the center line O-12 lies in the plane of the meridian, with 12 pointing north. Set the gnomon on the meridian at the corresponding date; then the position of the shadow of the

16th Century Horizontal Dial by Johann Gebhart.

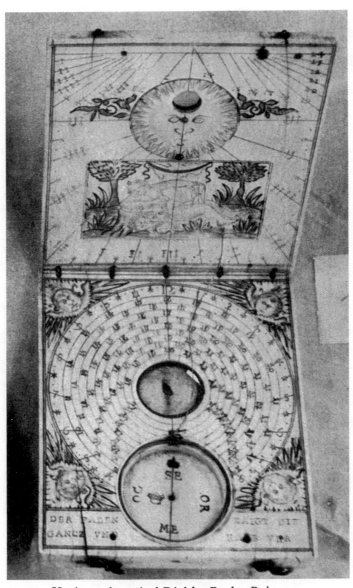

Horizontal-vertical Dial by Paulus Reinman.

gnomon on the ellipse will indicate the time. The time is shown only on the ellipse.

The practical side of this dial can be entertaining, when made to large scale on the ground, outdoors. You can use a tall rod as a gnomon, but it will be more fun if you stand on the point corresponding to the date. Your own shadow will indicate the time. To lay out a large dial, it will be easier to compute the points. Referring to (Fig. b),

Let

$n = 1/2$ major axis of the ellipse
$m = 1/2$ minor axis of the ellipse
$a = $ eccentric
$R+, R— = $ gnomon placements for north and south solar declinations
$H = $ hour angle
$L = $ latitude
$D = $ declination of sun ($+$ or $—$)

If n is known:

$$(1) \quad a = n \cos L$$

If a is known:

$$(2) \quad n = \frac{a}{\cos L}$$

or

$$(3) \quad n = a \sec L$$
$$(4) \quad m = a \tan L$$

To obtain the location of the hour points, we must use two coordinates: the distance u east or west of the meridian, and the distance t north or south of the major axis or 6 o'clock line.

Then

$$(5) \quad u = a \sec L \sin H$$

or by substituting formula (3) above

$$(6) \quad u = n \sin H$$
$$(7) \quad t = a \tan L \cos H$$

or by substituting formula (4) above

$$(8) \quad t = m \cos H$$

For the daily position of the gnomon on the meridian north $(R+)$ and south $(R—)$ of the major axis or 6 o'clock line:

$$(9) \quad R\pm = a \tan D$$

THE LAMBERT DIAL

This is an intriguing form of the analemmatic-type dial. It is a horizontal azimuth dial with a moveable gnomon or style, but unlike our previous example (Plate XXIX) the hour points are equidistant on the circumference of a circle. Also the gnomon makes an angle with the dial plate.

Construction (Plate XXX)

Draw lines WX and YZ, in (Fig. a), perpendicular to each other, intersecting at O.

From any convenient point E, on WX, draw a line making an angle G with WX equal to 1/2 the sum of 90° and the latitude of the place, and intersecting YZ at 12.

With O as a center and radius O-12, describe a circle. Divide the circle into parts of 15° each; these represent the hours.

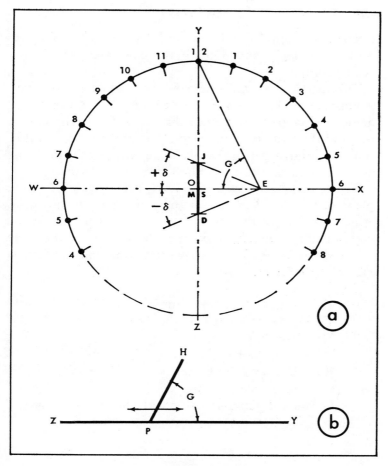

Plate XXX.

Lay a protractor on WX centered at E. Lay off angles of
the sun's declination for the days desired, as for the previous
dial. Draw lines from E through these points and mark their

intersection with YZ. The point J represents the summer solstice, while M and S mark the equinoxes, and D the winter solstice.

This dial does not use a perpendicular gnomon. However, it must lie in the plane of the meridian YZ and make an angle G (Fig. b) with the meridian equal to 1/2 the sum of 90° and the latitude of the place. The gnomon or style PH must be constructed so that its angle will remain the same as it is moved north and south along the meridian.

To use this dial, place it on a level surface with YZ in the plane of the meridian and the hour point 12 to the north. When the foot of the gnomon or style P is placed on the corresponding date, the position of the shadow on the circle will indicate the time. If you wish to compute the various elements, the formulae are:

Let

E = distance OE, the eccentric

C = radius of circle (distance O-12)

L = latitude of the place

D = declination of sun

R = distance from O north or south of WX on YZ, for the daily setting

G = angle gnomon makes with the meridian line YZ

Then

$$G = \frac{90° + L}{2}$$

$$R = E \tan D$$

$$C = E \tan G, \text{ or } C = E \cot (90° - G)$$

If the radius C is known, then:

$$E = C \tan (90° - G) \text{ or } E = C \cot G$$

XII

THE HELIOCHRONOMETER

Early sundials indicated local apparent time. We have shown you how the equation of time may be used to obtain local mean time from the apparent time of the sundial, and how the correction for longitude may be used to obtain standard time from the sundial (Chapter VI). Correction tables made for use with a sundial require some mental calculation. Many dials, both portable and stationary, were made in the 18th and 19th centuries that incorporated the equation of time in their construction. This enabled one to read local mean time directly from the dial. The analemma was the device that made direct reading possible.

When standard time came into use, a further correction was required in the dial's construction if a direct reading instrument was desired. One method was to incorporate the difference in longitude between the place and its standard time meridian into the analemma. Another was to use a standard analemma resulting from the plotting of the equation of time and a dial plate that could be adjusted for the difference in longitude. The latter method is the one we

193

shall describe, for it is the simplest and easiest to construct.

Description (Plate XXXI)

The heliochronometer is just what its name implies—a solar chronometer. Its accuracy is commensurate with its construction. It is not a toy but an instrument for education as well as for timekeeping. We shall use an equatorial dial plate which faces north and lies parallel to the equator.

The heliochronometer consists of four basic parts: base, dial plate, alidade or sighting instrument, and analemma.

The base is fixed in place.

The dial plate is attached to the base so that it may be rotated about its center.

The alidade is attached to the dial plate so that it can be rotated about its center, which is coincident with the center of the dial plate. Consisting of a flat plate, the alidade has two fixed upright arms perpendicular to the dial plate. One arm contains the style or nodus, the other the analemma.

The analemma here is based on the equation of time, when $E = A - M$ (same as the table on page 87 but with reversed signs).

The size of the dial is left for you to decide. We show the construction of the basic elements, and the diagrams are pretty much self-explanatory.

Construction (Plate XXXI)

The base WXYZ (Fig. a) may be a block of wood cut as shown in Fig. b, the angle WZH equal to the co-latitude (90° minus latitude). The bottom of the base is HZ, which must be leveled on a pedestal or other firm support and fixed

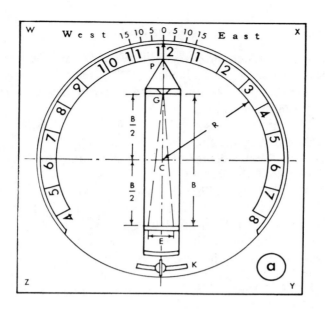

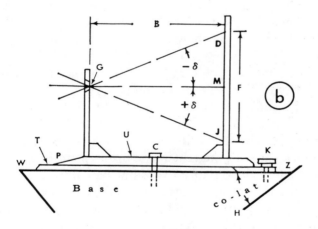

Plate XXXI.

in place. When properly adjusted, WZ will lie parallel to the equator; a perpendicular to WZ, erected at C, will point to the celestial pole; the center line of the face will lie in the plane of the meridian.

Here the base is square and represented by WXYZ. Bisect WX and WZ, and draw lines parallel to the sides, intersecting at C, the exact center. The dial plate and the alidade will rotate about the center C.

The dial plate (Fig. a) is shown as you would see it from directly above. With C as a center and radius R, describe a circle, which is then divided into equal segments of 15° for each of the hours. The number of subdivisions of each hour will depend on the size of the dial. A dial plate having a radius of 6 inches or more may be divided into minutes. The 12 o'clock line, if extended, would intersect the celestial equator at the true south point.

When you have determined the outside diameter of the dial plate, mark a point on the face of the base opposite 12 and label it o, as shown. Then with C as the center and radius Co, describe a 15° arc on each side of o and divide both at intervals of 1°. Mark these divisions East and West as shown. This will be the longitude scale.

The alidade must be proportioned to the dial plate, and centered at C. One end is pointed at P, which indicates the time. The center line of the alidade is its meridian and is coincident with the center line or meridian of the dial plate and base. The distance CG is one-half B.

The width of the arms is determined by the overall width of the analemma, E, which never exceeds plus or minus 16 minutes. Therefore, to be sure to have enough room, provide for 20 minutes or 5° on each side of the center line.

The height of the arms is determined in a similar manner, as shown in (Fig. b), by drawing GM parallel to the dial plate and laying off from G the declination of the sun at the winter and summer solstices. Lines drawn from G through these points will intersect the other arm at D and J. The point G is the style or nodus, and it is located in the south arm. The style or nodus may be a simple pinhole, the intersection of two cross hairs, or a bead centered inside of a small hole.

The line GM represents the path of a ray of light when the sun is on the equator at the equinoxes in March and September; GD at the winter solstice in December; and GJ at the summer solstice in June. The distance B is the same as in (Fig. a), and the distance F is the limit of the path a ray of light will travel during the year. The arms of the alidade must be kept perpendicular to the dial plate T at all times. All that remains to construct is the analemma.

The Analemma (Plate XXXII)

In (Fig. a) the point G represents the style; B the distance between the arms; D, M, and J the positions of the equinoxes and solstices; $\pm\delta$ north and south declinations; $\pm E$ the equation of time; GT the meridian of the dial and the analemma; GM the equator perpendicular to GT. Draw lines through D, M, and J parallel to GM.

Set a protractor on GT centered at G and lay off on each side of GT angles equal to minutes of time — $1° = 4$ minutes, $2° = 8$ minutes — and draw lines from G through these points to the solstitial line J. We have shown here only the lines representing 10 and 20 minutes ($2 \ 1/2°$ and $5°$).

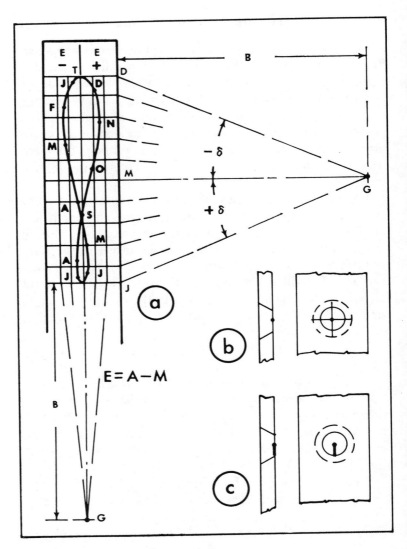

Plate XXXII.

The positive and negative equations must be marked as shown when E = A — M.

Now set a protractor on GM centered at G and lay off on each side of GM angles at 1° intervals. Draw lines from G through these points to the side of the arm DMJ, and through the points on DMJ draw lines parallel to GM. We have shown here only 5° intervals and the declination of the sun at the solstices. North and south declinations must be marked as shown.

This provides you with a grid upon which the equation of time may be plotted easily. The vertical lines are minutes of time, the horizontal lines degrees of declination. Plot as many points as necessary to obtain a smooth curve: the 1st and 15th of each month, the points where the equation is greatest and where it is zero, and the equinoxes and the solstices. As many days as desired can be marked on the curve of the analemma. This completes the dial, except for making a slot in the dial plate at K (Plate XXXI, Fig. a), and fitting it with a thumbscrew or other device by which the dial plate may be locked in place.

Using the Dial

When properly set, the alidade will be in the position shown on Plate XXXI (Fig. a) when it is noon local apparent time; and at four times during the year it will be in this position at noon local mean time.

To obtain local mean time at any hour on any day, just turn the alidade until the light passing through the hole at G is centered on that portion of the analemma corresponding to the date. In this position P will indicate local mean time.

To obtain standard time we must adjust the dial plate for the difference in longitude between the place and its standard time meridian. Unloosen the thumbscrew and turn the dial plate until the 12 o'clock line is opposite the number of degrees on the longitude scale equal to the difference between the meridian of the place and the longitude of the standard time meridian. Since this always remains the same, the dial plate may be locked in this position. Then, turn the alidade until the spot of light is centered on the analemma corresponding to the date in question. The pointer P will now indicate standard time.

You can make the dial more interesting if you mark every five days on the analemma and mark the degrees of declination on one side and the degrees of altitude on the other side of the analemma. With this data incorporated, the dial becomes more useful. You can use it to show the declination and altitude of the sun, the day of the year, apparent time, mean time, standard time, and the equation of time.

If you do not wish to use a dial plate that can be set for the difference in longitude, you can incorporate the difference in the analemma. If you are east of the standard meridian, subtract the difference from the equation of time; if west, add the difference. For example, assume a dial on each of three meridians reading $2^h 15^m$ p.m. on April 1. The equation of time is $+3^m 59^s$, or say $+4^m$. Since the equation here is based on $E = M - A$, the equation is added to the time of the sundial to obtain mean time. Then we can add or subtract the difference in longitude to or from the equation to obtain the total correction to be added to the reading of the dial to obtain standard time, as follows:

If the three dials mentioned above are on the 77th, 75th,

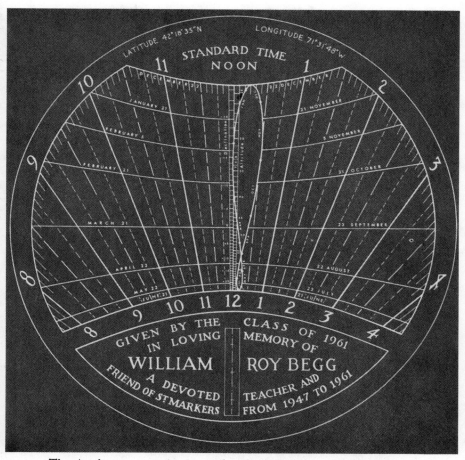

The Analemma permits Local (sundial) Time to be corrected to
Standard (watch) Time.

A Wall-type Dial at Rapperswil, Switzerland.

and 72nd meridians, the computation for the corrected analemma is:

77th meridian $= 2°$ long. or 8^m west

$$\text{Cor.} = + 4^m + 8^m = + 12^m$$

75th meridian $= 0°$ long. or 0^m

$$\text{Cor.} = + 4^m + 0^m = + 4^m$$

72nd meridian $= 3°$ long. or 12^m east

$$\text{Cor.} = + 4^m - 12^m = - 8^m$$

The result is the hour angle for the corrected analemma, which is plotted in the same manner as for the standard analemma; but first reverse the $+$ and $-$ signs for E, Plate XXXII, Fig. a. When the corrected analemma is plotted with respect to the meridian of the alidade, the dial plate is set with the 12 o'clock line always opposite 0 on the longitude scale, or in the plane of the meridian.

If you want to make a more accurate dial than can be done by geometric construction, the shape of the analemma may be computed.

Formulas for the Analemma

Trigonometric tables are available with the angles given in hours and minutes. If you do not have such a table, you must first convert the equation of time into degrees and minutes of arc for the number of points desired.

Let

$\quad$ E $=$ horizontal distance from meridian GT, Plate XXXII, Fig. a

$\quad$ D $=$ vertical distance from equator GM, Plate XXXII, Fig. a

B = distance between arms of alidade, Plate XXXI,
    Fig. b
H = equation of time in hour angle or in degrees
δ = declination of sun

NOTE: On Plate XXXI, Fig. b, the style G and the
      equator M must be equidistant above the dial
      plate.

Then

$$D = B \tan \delta$$
$$E = B \tan H$$

XIII

SUNDIAL CLASSIFICATION

W<small>E HAVE</small> pointed out the great variety of forms, materials, and methods of telling time, used in the construction of portable dials. Obviously there are many more portable sundials than stationary ones. If you look over a large collection of portable dials certain discrepancies become apparent. For example, with so many different looking instruments displayed in one place, the obvious question is, "What kind of a dial is it?" That question leads to others: "How is it used? How does it tell time? and What tells the time?" The first question is the most important.

Let's take a concrete example. While looking over a catalog of a large collection, six instruments attracted our attention. They were listed as follows: 1, Horizontal Dial; 2, Compass Dial; 3, Tablet Dial; 4, Pocket Dial; 5, Octagonal Dial; 6, Ivory Dial. However, each piece is placed in a horizontal position when in use; all tell time in exactly the same manner; all are made for use in particular places; and all have compasses. One might expect six different types of dials, but they differ only in form, material, and maker. They are all horizontal dials.

Similarly you will find the following notations: 1, Ring

Dial; 2, Universal Dial; 3, Equatorial Dial. Here again these three dials all have the same form and method of telling time. They are equatorial dials and tell time by a ray of light. In other words, in the previous example the dials must be laid flat so that the dial plate is horizontal; and in the second group the dial plate must be parallel to the equator. Obviously such notations regarding dials are very confusing to the student.

After examining several thousand portable dials it became apparent that a dial should be classified by the position of the dial plate when in use. There are a few instances where the form is such that a more descriptive term is needed to identify a special type. The armillary, for example, has a very definite form, although the dial plate is "equatorial"; that is, it lies parallel to the plane of the equator.

Classifying pieces in any collection requires some knowledge of the subject concerned. In the case of sundials you should be familiar with certain parts, and the principle upon which the instrument works. The fundamental parts, common to all dials whether stationary or portable, are *style, gnomon, dial plate, substyle,* and *nodus.*

In order to determine the type of a sundial, it is necessary to know only the position in which the dial plate lies when in use and the form or appearance of special dials like the armillary. Basic type nomenclature, descriptive of the position or form, is used in the following discussion of various categories in which all dials can be placed. The special group comprises those dials which require detailed explanation or specific definition. The accompanying chart lists these basic and special types in full capitals.

Ordinary Types

HORIZONTAL. The most prevalent of dials. May have multiple hour bands, a fixed or folding gnomon, solid or thread style. Usually fitted with a compass (portable form). For universal use, fitted with fixed style, quadrant, and single hour band.

VERTICAL. These dials are upright and face the cardinal points of the compass. Usually single hour band with fixed style. On north and south dials the substyle lies in the plane of the meridian (12 o'clock line). The east and west dials have parallel hour lines and a fixed style elevated above and parallel to the dial plate. Substyle is the 6 o'clock line.

DECLINING. Vertical dials that do not face the cardinal points of the compass. The substyle is not the 12 o'clock line (meridian).

RECLINING. Usually found on multiface dials (see below) and often used on sloping roofs, wall copings, and so forth. They face the cardinal points of the compass and lean from you as you look at them. The polar and equatorial (see below) are reclining dials, but they are separated and named in accordance with the plane in which they lie. The north and south recliners have the substyle on the 12 o'clock line. The east and west recliners have this line horizontal, and the substyle does not lie along it.

INCLINING. Usually found on multiface dials. These are similar to the reclining dials but lean toward you as you look at them. They face the cardinal points of the compass.

DECLINING-RECLINING. These are neither vertical nor do they face the cardinal points. The 12 o'clock line is not perpendicular and the substyle is not the 12 o'clock line. Usually found on multiface dials, but also used on roofs, wall copings, and so forth.

DECLINING-INCLINING. These are the opposites of the declining-reclining dials and are similarly identified. Usually found on multiface dials.

EQUATORIAL. The hour lines are equally spaced except in certain cases such as those dials in the form of a star or a cross. Other common forms include concave and convex half-cylinders or bands, full and split rings, convex and concave hemispheres, and globes. For universal use, fitted with a quadrant. May be inscribed on upper and lower surfaces of the flat plane variety. In the case of rings, either a double-needle gnomon is used or a single reversible needle, or a pinhole sight. Some—particularly standard-time dials and heliochronometers—are fitted with an alidade (sighting device). See also RECLINING above.

POLAR. The hour lines are parallel. Fixed style elevated above and parallel to dial plate. For universal use, fitted with quadrant. Usually found on multiface dials. Dial plate lies parallel to axis of the earth. See also RECLINING on the previous page.

MULTIFACE. These dials comprise only those solids (or hollow solids) with two or more faces inscribed with hour lines. A common form is the cube, which is also made for universal use (fitted with quadrant and plumb). Combination dials such as horizontal-polar-equatorial, or horizontal-vertical-equatorial, belong in this class and may be found among stationary dials.

Special Types

HORIZONTAL-VERTICAL. A definite type of portable dial, usually in the form of two hinged tablets. The horizontal dial plate often has multiple hour bands. Solid or thread gnomon. For universal use, fitted with quadrant and single hour band. The universal type is rare, but was recently brought into use in exploration. Stationary dials of this type are few.

ARMILLARY. A definite form comprising a system of rings (from 2 to 10), corresponding to the major circles of the terrestrial and celestial spheres. The axis of the sphere is the style. Hour lines are equally spaced on the equatorial band or ring. For universal use, the meridian ring is graduated in degrees. Often a ray of light serves as an indicator, passing through a pierced slide that may be adjusted for the day of the year.

ANALEMMATIC. A specific type of horizontal dial comprising a combination of HORIZONTAL and HORI-ZONTAL (Azimuth), see below. Easily distinguished by the elliptical form of the hour band and the perpendicular style which moves in a north-south direction on the meridian or minor axis of the ellipse. The style (here coincident with the gnomon) is set to the corresponding day of the year. If the elliptical band is a separate dial by itself, the class is then HORIZONTAL (Azimuth).

SIGNAL GUN. Often combined with a horizontal dial. A lens is used to focus the sun's rays on the touch hole of a cannon which is so placed that the charge is set off at noon, apparent time. The lens is adjustable to the declination of the sun.

CONICAL. Comprises all dials in the form of a cone, either partial or full. The so-called goblet or chalice dials are conical dials. They tell time by recording the altitude of the sun, and they are easily distinguished. A vertical pin in the center, or the edge of the rim, may be used to cast the shadow.

CONCAVE. This type includes those dials inscribed on concave surfaces that cannot be relegated to one of the foregoing types. Many equatorial dials are concave but the position of the dial plate determines the type. The ancient hemisphere of Berosus is typical of the concave type, as are many of the "sunk" dials of the Renaissance.

CONVEX. Includes dials on convex surfaces that cannot be categorized above. Many equatorial dials are convex.

Such are the so-called basic types. All dials can be assigned to one type or another, but it is a general classification and does not definitely identify an individual work. In a group of vertical dials, some may be flat, others round; the hour lines may be straight or curved; and the means by which the time is recorded may be different. Therefore, certain additional data are necessary to define clearly an individual piece. This is provided for in the classification by referring to three things: *use, method,* and *indicator.*

Use. Some dials are to be used in a particular place or places. Such dials are easily recognized by the single hour band, for a particular location, or multiple bands, usually three, four, or five. In the case of multiple bands, each is labeled by a figure noting the latitude for which it is to be used. Dials that function anywhere are called *universal* dials. They are usually fitted with a quadrant and plumb, or some other means of adjusting the dial plate to any latitude. Then there are dials designed for *specific* purposes, such as the noon mark, which is used to record only the noon hour.

Thus a HORIZONTAL dial is just a plain dial to be used in one or a few specified places; or it may be a HORIZONTAL, Universal, for use anywhere; or it may be a HORIZONTAL, Noon Mark, to determine midday.

Method. This refers to how the dial tells time—by measuring the *hour angle, altitude,* or *azimuth* of the sun. The familiar horizontal dial tells time by recording the hour angle of the sun; that is, the angle (measured in time) on the celestial equator between the sun and the true south point of the celestial equator. This method is most common.

Many interesting dials tell time by measuring the altitude of the sun (angular height above the horizon). Typical is the cylindrical dial, with a gnomon which projects outward from and at right angles to the side of the cylinder. This cylinder is marked with vertical date lines (over which the gnomon must be placed by turning the knob at the top), cut by curved hour lines. Similar dials will be found in the form of disks, quadrants, or rings. Because the portable kinds are usually suspended when in use, they are called VERTI-CAL (Altitude). Regardless of the position in which they lie, they may be recognized generally by the oddly shaped hour lines cutting across date lines. The point of the gnomon is the style.

The angular distance, measured on the horizon, between the south point and the foot of the perpendicular from the sun to the horizon is called the *azimuth*. Dials that use this method to record the time are called azimuth dials. Like altitude dials, they can be recognized generally by wavy hour lines crossing date lines. The gnomon is usually a pin.

Indicator.   The device used to indicate the time is called the indicator, and may be shade, light, or a magnetic needle. The first two are obvious, but the latter is not so easily determined. Dials of the magnetic type are rare and look as though the gnomon had been lost. They would, of course, be classified as HORIZONTAL (Azimuth) Magnetic. The hour lines are wavy and cut by date lines. A magnetic needle is in the center. The whole is mounted in a square block which, when turned until the east and west sides cast no shadow, will allow the needle to indicate the time.

If there can be glamour in sundials, it will be found in those that have other lines and parts in addition to the gnomon, hour lines, and style. All such parts or lines not essential to the primary function of the dial are called *furniture.*

On stationary dials, the common furniture includes lines of declination, signs of the zodiac, an analemma, lines of altitude and azimuth. But on the portable dials the common furniture includes all of the foregoing together with the compass, windrose and vane, lunar calculators, nocturnal, Babylonian hours, Jewish hours, and perpetual calendars of various sorts.

Lines of declination, altitude, and azimuth can be deciphered by a casual inspection of the dial and gnomon. They are usually numbered in degrees and/or days. The days usually refer to the equinoxes and the solstices, but frequently the 20th or 21st of each month is noted. Lunar calculators are of two kinds—those that require a previous knowledge of the age of the moon, and those by which the age of the moon can be determined (often called lunar phase dials).

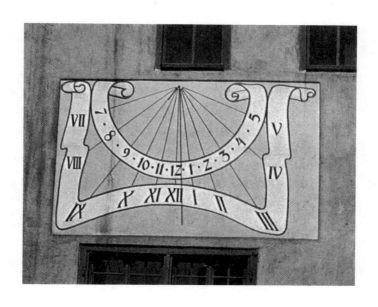

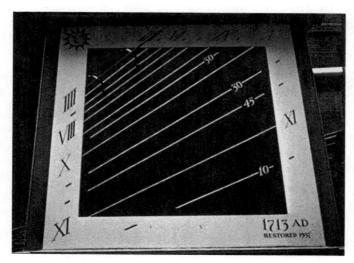

1713 AD
RESTORED 1957

NÖRDLINGEN, GERMANY

The light-colored dial face contrasts nicely with the wall color of the "Red House." The dial itself is 8' by 10' in size.

BOSTON, MASSACHUSETTS

In 1957, R. Newton Mayall restored this dial on the facade of the Old State House in Boston. Note that only morning hours are given, since the dial is located on an eastward facing wall. The Arabic numerals indicate fractions of hours, an especially important innovation toward midday, when the shadow of the gnomon (top left) moves very rapidly across the dial face.

The calculator is used to determine the time, without mental arithmetic, by observing the shadow cast by moonlight on the ordinary sundial. Either type will have a circle divided into 29 or 29 1/2 parts.

The nocturnal is not a sundial, but a portable or stationary dial used to tell time at night by means of the stars. It consists of a base plate or disk, bearing the days of the year, on which a rotating hour disk and an alidade are mounted. A hole is made in the center. The nocturnal is usually designed for use with the pole star and one or both of the Pointers in the Big Dipper, or the pole star and the brightest star in the bowl of the Little Dipper. The hour disk may be notched to permit reading the time in the dark by counting the number of notches between the reference point of 12 and the sighting edge of the alidade. The instrument is frequently incorporated in portable sundials.

The Babylonian, Italian, and Jewish hours may be recognized by their appearance and the accompanying numerals. The Babylonians counted their hours continuously from one to 24 from sunrise to sunrise; the Italians counted theirs continuously from one to 24 from sunset to sunset. Thus the Babylonian hours are noted from 24 to 12, whereas the Italian hours are noted from 12 to 24. The Jewish hours are frequently referred to as the "old unequal planetary hours" or as the "temporary" hours. This arises from the division of the day from sunrise to sunset into 12 equal parts. Because this period varies in length, the summer daylight hours will be of longer duration than the winter hours. Similarly, the hours of night will be just the reverse—longer in winter than in summer.

Calendars are of many different kinds serving many

different purposes—from fixed annual to perpetual and from simple to complex. Some enable one to determine for any given year the golden number, dominical letter, epact, cycle of the sun, the time of new moon, and so forth.

The analemma is a device used to make a sundial record local mean time, or standard time. It usually appears in one or two forms, as a figure eight, or linear (in the form of a scale). Sometimes a two-dimensional chart or a table is appended to the ordinary dial. In other dials, the analemma is incorporated in the hour lines, causing a deformation, in which case it is not classed as furniture.

Furniture is most prevalent on European dials. Oriental dials seldom have furniture, except for the compass. Lunar calculators and calendars will be found occasionally on Oriental dials.

KEY TO CLASSIFYING, LABELING, AND CATALOGING SUNDIALS

OUTLINE FOR LABEL
(Numbers Key to Divisions Listed Below)

0	1	2	3	4	5	6
No.	TYPE ,	Use	(Method)	Indicator -	Country -	Date

1 - LIST OF TYPES

Ordinary	Special
HORIZONTAL	HORIZONTAL-VERTICAL
VERTICAL	ARMILLARY
DECLINING	ANALEMMATIC
RECLINING	SIGNAL GUN
INCLINING	CONICAL
DECLINING-RECLINING	CONCAVE
DECLINING-INCLINING	CONVEX
EQUATORIAL	
POLAR	
MULTIFACE	

NOTE: Use capital letters for division 1. Use capital
       and small letters for divisions 2, 3, 4, and 5.

2 - USE	3 - (METHOD)	4 - INDICATOR
a - Particular place or places (omit)	a - Hour angle (omit)	a - Shade (omit)
b - Universal	b - Altitude	b - Light
c - Specific (such as "Noon Mark")	c - Azimuth	c - Magnetic

COMMON FURNITURE

Compass	Babylonian Hours (1-24, Sunrise to Sunrise)
Windrose and Vane	
Lines of Declination	Italian Hours (1-24, Sunset to Sunset)
Lines of Altitude	
Lines of Azimuth	Jewish Hours ("unequal" 1-12 Sunrise to Sunset
Signs of Zodiac	and Sunset to Sunrise.
Lunar Calculator (two types)	Also on Oriental dials)
Nocturnal	Calendars
	Analemma (various forms)

56	ARMILLARY	Universal	—	Light
No.	TYPE	Use	(Method)	Indicator

Germany
Country

ca. 1720
Date

Maker   Johann Willebrand

Material   Brass

Size   85 mm diam.

Description:-

2 Rings - hour band equatorial. Both rings engraved with names of cities and their latitudes. Meridional ring divided in degrees in one quadrant only. Pierced slide in polar axis adjustable to day of year. Collapsible with leather case. Meridional ring inscribed "Johann Willebrand in Augsburg".

GNOMONIC CARD

Copyright 1971 by R. Newton Mayall

Sample Gnomonic card based on a standard library index card.

XIV

INTERESTING DIALS OF THE WORLD

The Great Dial at Jaipur

THE United States can boast of many largest things in the
world, such as office buildings, tallest buildings and so forth,
but the largest sundial in the world is at Jaipur, India. At
least we know of none in existence or contemplated that
is larger. Jaipur can also boast of the largest collection of
large stationary dials in the world.

The Great Dial at Jaipur alone occupies nearly an acre of
ground. Its sloping gnomon is well over one hundred feet
long and wide enough to allow for steps that one may climb
up the sloping surface to a covered observatory at the top
(facing page 216), where he may look out upon the coun-
tryside, the other instruments scattered about over a great
area, see illustration opposite page 152, or he may watch the
shadow move across the curved surface of the dial.

The dials were constructed about 1724 by Jai Singh, Rajah
of Jaipur. The shadow casting edges of the gnomons are
marble as are the dial faces. Words cannot adequately describe this great display. The illustrations tell their own
story, which with the plan of the grounds, see Figure 31, may

give some idea as to the gigantic size of the observatory and its instruments.

For many years the grounds were neglected and many of the dials had crumbled to pieces as can be seen in the illustration opposite page 152. Fortunately, this great observatory was restored about 1901.

A DESIRABLE INSTRUMENT

In sharp contrast with the great dials at Jaipur is the small portable instrument, see illustration opposite page 153, which is only 7″ long, 2 1/2″ wide, and 1″ high. It was made in Japan of olive wood and contains in addition to the small brass hemisphere, an inkwell, compass, abacus or counting apparatus, brush for writing, a small pair of scissors, two ivory-handled drills or needles, and a knife. Could anything be more compact or useful. This and several other similar dials of various forms are in the Ernst Collection.

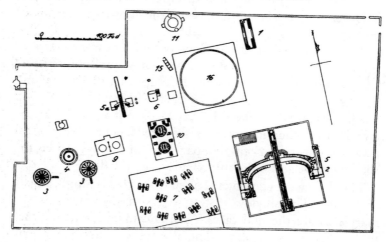

Fig. 31.
Plan of observatory at Jaipur, India.

The Largest Sundial in the World, at Jaipur, India.

Dial at Portal of the Church of Brou, Bourg, France.

A Noon Mark Goes to Sea

How many of you know that sundials were used at sea? The illustration opposite page 153 shows such a device in the Ernst Collection, mounted on gimbals with a heavy weight suspended beneath the plate to keep it level. This is a brass Japanese dial which records only the noon hour. The plate is fitted with a compass, to aid in setting. It may be dismantled when not in use and carried in the wood box, which appears in the illustration.

An Armillary

One of the finest armillaries in the United States is situated on the campus of Phillips Academy, in Andover, Massachusetts. See illustration opposite page 160. We are indebted to Dr. Claude W. Fuess, Headmaster of the Academy, who kindly furnished the accompanying photograph and the following description, written by Mr. Paul Manship, the designer and sculptor of this unique sundial.

"The path of the sun is shown by the Ecliptic and the Signs of the Zodiac are portrayed in high relief on the band of the equator. The shaft, representing the axis of the earth, points to the North Star; and its shadow on the belt of the equator indicates the hour. The four Elements, as well as Dawn and Evening, figure in the decorative scheme: Water in the wave motif, with the Earth motif growing out of it; Air is represented by the ribbon, and Fire on the flaming meridian. The whole is supported by turtles, emblems of eternity. Man, Woman, and Child make up the Cycle of Life, as the sphere itself symbolizes the Cycle of Eternity."

This sphere with its pedestal and base, as a unit, might well be symbolical of character, strength, and dignity, thus being an ever-present reminder of the responsibility of all such institutions to men in the making.

The Whitehall Dial

A Royal dial is that shown in the illustration opposite page 166, set up in the "King's Majesty's Privy Garden at White-Hall in 1669." It is an excellent example of the grotesque and the length to which men went to gain favor with their king.

The dial stands about ten feet high and is made up of six different parts. The first or lower part is a round table about 40 inches in diameter, with 20 dials arranged around its edge. The dials are glass, set into the table; some show the hours in accordance with the ancient manner of the Jews; some according to the Italian method of reckoning; others the time used by astronomers; and still others the time of everyday life.

The reclining dials on top of the table are also covered with glass and show the time in several ways—by the shadow of a style falling on the hour lines; by the shadow of the hour lines falling on the style, etc. The globes supporting the second part contain dials relating to astronomy, geography, planetary motions, etc.; and on either side may be seen glass bowls supported by brackets which point toward the cardinal points of the compass. The illustration shows only half the true number of dials.

There are sixteen dials around the edge of the second piece, like those in the first piece. They differ from the former in that the dials are not laid out on glass but are drawn on the bottoms of small boxes cut into the table and covered with glass. Neither do these dials tell the time of day, in the usual manner, but rather by the rising of various well known stars.

On the top of this second piece are 8 inclining dials; four of them are reflected to the sloping surfaces at the bottom of the third piece, whereas the other 4 may be seen in mirrors placed on the inclined surfaces.

The third piece rests on the second. It is cut into 26 faces, each containing a dial. The fourth piece is another table-like affair, with its edge cut into 12 equal surfaces containing concave dials in the form of cylinders. The fifth piece is a polyhedron containing 12 triangular faces and 6 square faces on which are displayed the usual hours.

The sixth or last piece, atop the pyramid, is a glass bowl about 7" in diameter. The north side is thinly painted white so that the shadow of a small gold ball will point out the hours on the white surface. The whole is surmounted by a cross.

The Whitehall dial had but a short life, for by 1700 it had been demolished. It was made by the Reverend Father Francis Hall, and was set up for King Charles II. Father Hall wrote a book describing the various parts in which he states that among the "very many dials, especially the most curious, are new inventions hitherto divulged to none. All these particularly are shortly yet clearly set forth for the common good."

FLOWER TIME

All garden lovers will be interested in the floral sundial in Cypress Lawn Memorial Park, San Francisco, California, shown in the illustration facing page 167. It is made entirely of flowers as is the motto above it. Such dials are rare in the

United States, but they are not uncommon in England where clipped yew or box is used for the gnomon and the dial plate laid out in a variety of schemes.

A floral sundial makes an excellent feature in any public park or cemetery. Due to the character of materials it must be designed on a large scale and the position of the hour lines should be computed mathematically for use with a surveyor's transit. Care must also be used in the selection of plants.

The illustration shows a dial ably and beautifully executed, second to none. Mr. Noble Johnson of Cypress Lawn Memorial Park generously supplied the photograph together with the following description:

"The dial, itself, is fifty feet in diameter and is made entirely of growing plants. The gnomon is made from a Cypress tree and it is entirely covered with growing Ivy that is kept closely trimmed. The numerals are planted with Santalina, a gray close growing plant. The field and borders are planted with fibrous begonias (Luminosa compacta), Acaranthus, Iresine and yellow Pyrethrum (Carpet of Gold)."

A Moon Dial

Many a traveler has stopped to look at the famous moon dial at Queens' College, England, painted on the masonry wall. The three rows of figures, below the dial, see illustration opposite page 42, have often caused a great deal of speculation. They are, however, the secret of the moon dial, for without them no one could tell time by the light of the moon.

President Venn very kindly sent us the following description of the dial:

Singing Tower Dial, Lake Wales, Florida.

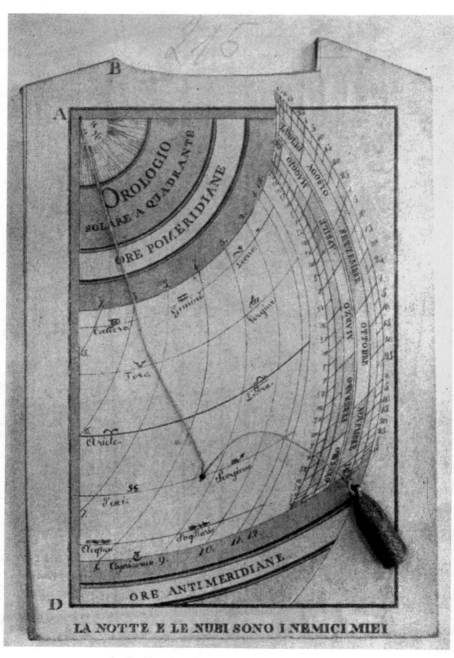

An Italian Quadrant Dial of a century ago.

"The real object of the extension is to enable the dial to play the part of a Moon-dial as well as that of a Sundial, in a manner which we must explain.

"If we could see traced out on the sky the path of the sun during a whole day, the moon would always be found in or close to that path; and the distance of the moon ahead of the sun would be simply proportional to the moon's age, a new moon being extremely near the sun (an eclipse of the sun can take place only when the moon is new, a fact of which not every writer of fiction seems to be aware), a full moon half a day's journey away, and the moon as it dies at the end of the lunar month overtaking the sun to commence again. We take the lunar month to consist of thirty days. Thus, for example, a five-days-old moon has completed one sixth of its monthly course and is therefore one sixth of a day's journey ahead of the sun in the sky. Suppose further that some wakeful inhabitant of the Old Court sees that the moon, five days old, is casting the shadow of the style across the hour-mark IX, he knows that in one sixth of a day, that is four hours time, the sun will reach the place in the sky now occupied by the moon, and will therefore cast the shadow of the style across the Figure IX. That is, in four hours time it will be nine o'clock; it is now five.

"But what, the reader asks, has the mysterious table to do with this? Indeed the part it plays is quite trivial I fear. It saves us the labour of calculating from the moon's age the number of hours and minutes by which moon time is in advance of clock time; and since an addition or subtraction of a round twelve hours, that is, half a day, makes no difference to clock time, the addition or subtraction from the moon's age of half a lunar month, that is, of fifteen days, leaves unaltered the amount by which the moon exceeds clock time. The entries for the second half of the month would therefore be a repetition of those for the first half; and so, instead of actually repeating the figures, the designer of the table has shown the two ages to which the same entry applies, putting the number of hours and minutes in excess in the second row while the corresponding ages of the moon in the first half of the month are in the first row and the corresponding ages of the dying moon are in the third row. Thus the process of using the dial as a moon dial is as follows. First ascertain roughly the age of the moon; this can be told with sufficient accuracy for our present purpose, by mere notice of the phase of the moon; the first half moon is 7½ days old, the full moon 15, and the second half moon 22½, but the phase is changing more rapidly when about half the disc is illuminated, so that the moon is already 5 days old when the breadth of the crescent is only a quarter of the total diameter, and is only 10 days old when three-quarters of the face is bright, similar remarks applying also to the phases in the second half of the lunar month. Now look in the first or third row of the table for the entry nearest to the estimated age of the moon, and the corresponding figures in the second row give the number of hours and minutes by which the reading given by the shadow of the style is in advance of clock time. Subtracting then the second row reading from the shadow reading, having first added twelve hours to the latter if it does not exceed the former, we find the actual time, by a process interesting perhaps but certainly not very accurate, for not only would an error of less than a day in judgment of the moon's age be sufficient to modify the result to the extent of three-quarters of an hour, but also the motion of the moon. . . . The reader who infers from the moonlit dial a time differing by less than an hour from that announced by the clock above it, will have every reason to be satisfied with his performance."

From the above it seems quite a chore to deduce the time by the moon. However, we know that many who have seen the dial will be glad to read the explanation.

The dial plate is a striking thing in itself, for the outside border containing the hour numerals is blue, the sun at the center is golden, the vertical panels on either side contain the signs of the zodiac, and the curved lines crossing those of the hours show the declination of the sun and the day of the year. The vertical lines show the position of the sun with respect to the points of the compass. The dial was painted on the wall in 1733.

CATHEDRAL DIALS

Throughout Europe and Great Britain many dials of one kind or another are found on cathedrals and small churches. There is hardly a city or town in some localities where at least one can not be seen. Innumerable books contain many pictures of them. It would be folly to say that any lack interest, but we have selected two dials of extreme interest that are not well known. One is in France, see illustration opposite page 217; the other is in the United States, see illustration facing page 244. They are of interest, because of their character and use, and neither is attached to the cathedral building.

The unusual dial at the portal of the Church of Brou at Bourg, France, was brought to our attention by Roland W. Taylor, who has graciously allowed us to reproduce his photographs. It will be recognized as an analemmatic dial, but it goes further than that—the equation of time is incorporated in the figure eight carved in the oblong stone situated in the center of an ellipse outlined with small pieces of marble. If the perpendicular iron gnomon is placed over the mark on the figure eight which corresponds to the proper day, the shadow cast will point toward the hour of standard time for the locality.

The position of the hours is marked by larger pieces of marble with numerals cut on the top. This dial was laid out in 1756, and is about 35 feet in its largest dimension.

Our second cathedral dial is known as the "Cathedral Landmark", situated on the grounds of the Washington

Cathedral, Mount Saint Albans, Washington, D.C., see illustration opposite page 244. Its inclusion here was made possible by the generous cooperation and help of Mr. Herald L. Stendel, in time of need. The photographs were taken especially for this book.

Mr. Stendel writes:—"The following paragraph appears in our guide book—

"Landmark. The Cathedral Landmark, a few feet north of the Thorn, commemorates the freeing of the Cathedral land from debt and the subsequent hallowing of the Cathedral Close. The donor of the last $50,000, Mrs. Julian James of Washington, set up on Ascension Day, 1906, this beautiful bronze sundial on which are inscribed the names of those it commemorates. The sundial marks not only the hour of the day, but the different seasons of the Christian year."

Continuing he describes the dial thus—"The 'Cathedral Landmark' is a kind of sarcophagus of limestone which rests upon three granite steps. Atop the sarcophagus is a bronze plaque six feet long and four feet wide. In relief upon the plaque is a large Latin cross, upon the cross of which is inscribed the hour circle for a horizontal sundial.

"Near the north edge of the hour circle and on the stem of the cross stands a vertical Latin cross also of bronze. At intervals along the stem of the cross in relief are marked the religious seasons—Epiphany, Lent, Easter, Ascension, Whitsunday, St. Peter—St. Paul, Transfiguration, Michaelmas, All Saints, Advent and Christmas.

"The top of the vertical cross casts a shadow, which, at noon, points out the religious seasons. The complete dial is now covered with a rather light green patina."

What more fitting memorial could be placed on hallowed ground.

There is little more that can be said about this beautiful dial. It is not a cruciform or cross dial because the arms of the cross do not indicate the hours. It is really an altitude dial which is based, in this case, on the declination of the sun, at noon, apparent time. Since the sun's angular height above the horizon is known for each day in the year, the shadow of the cross can and will unerringly note the days, or the religious seasons as on the Cathedral Landmark.

Singing Tower Dial

The dial on the Singing Tower at Lake Wales, Florida, should not be passed by. See illustration opposite page 220. It is a beautiful dial in a beautiful setting. Beneath the dial is a chart by which standard time can be obtained by applying the proper correction to the reading of the dial. If you have forgotten what day it is, a glance at the position of the shadow of a specific point on the gnomon will tell you. The accompanying photograph is reproduced with the kind permission of the Architectural Record, in which magazine it appeared in August 1937, together with several drawings of the dial plate.

Exploration

From time to time the work of sundial enthusiasts comes to our attention. We have been much interested in one who has made several dials commemorating certain events of especial significance, such as the total eclipse of the sun which was visible over northern New England where hundreds of thousands witnessed the spectacle.

We speak of the Reverend Joseph R. Swain of Connecticut whose latest edition celebrates the polar expeditions of Admiral Richard Evelyn Byrd, see illustration facing page 244. In response to our request he has graciously consented to describe his own creation for you.

"The large panel presents the sun, symbolizing the long

polar day when expeditions operate in the field, the regions explored, and the polar plane. The small relief under the gnomon shows the leader of the expeditions using his sextant, and the panels right and left above the numerals show radio towers thus celebrating the means of communication which helped to make the Byrd Expeditions so effective in the field and so continuously interesting and thrilling to the folks back home. The motto: "By Endurance We Conquer", made famous in the annals of polar exploration by the British explorer Sir Ernest Shackleton whose leading sledge almost invariably carried this family emblem, has been given new meaning and splendor for Americans by the courage and endurance of Admiral Byrd and his companions. The gnomon symbolizes the polar night, incised circles on the gnomon present the stars of the northern and southern regions and the symbols of four of the great circumpolar constellations: Crux and Argo Navis for the South, Draco and Ursa Minor for the North."

Mr. Swain has ably incorporated in his sundial two great epochs in the life of a great man. Truly an interesting dial of the world, evoking world interest.

. . . And Fired the Shot Heard at Noon

The signal gun fires a shot at noon. We have watched staid adults step quickly back into childhood in the presence of this type of dial. However, it is not a toy. We are grateful to the Hamilton Watch Co. for permitting us to reproduce the finely wrought signal gun in their collection, see illustration opposite page 245. They describe it thus:

"Among the most interesting of all developments of the sun dial are those instruments which use the power of the sun not only to cast a shadow but to do a certain job.

"When the Hamilton Watch Company began its collection of various timekeepers, they decided to specialize on quality and uniqueness rather than quantity or variety. Consequently the sun dial selected by them is one of the rarest forms known—a cannon dial or sundial gun. The model in their possession was built by Rosseau of Paris about 1650. Upon a marble base is mounted a small brass cannon whose touch-hole has been elongated into a groove that exactly parallels the North and South line on the dial. Immediately above the cannon is mounted a burning glass lens which mounted upon struts, can be set for the various months of the year. When set for December, the glass is four inches lower than when set for June. This is necessary because the sun is much lower in the sky during the Winter than during the Summer. The little gun is loaded every day with approximately a teaspoonful of powder and the long touch-hole is sprinkled lightly with powder. A dry wadding is rammed home in the muzzle of the gun and when the dial is mounted upon the parallel for which it is cut, the gun would discharge at twelve o'clock noon fired by the concentrated rays of the sun as it crossed the line.

"The Sultan of Morocco owns a sundial of this type carefully made by Baker & Sons in London. Sundial guns may be found in several European towns. And they were sometimes used on shipboard. Very often the burning glass was simply mounted above the gun set on a swivel. This was necessary on shipboard due to the fact that the gun would have to be set due North and South by the ship's compass. Thus the gun fired approximately at noon and was often known as the noonday gun.

"The invention of the first successful ship chronometer practically eliminated the use of the sundial gun on any but third rate vessels. Notwithstanding the fact that their usefulness has passed, the sundial guns are most interesting mementos of the inventive genius of a past generation. Invariably these replicas draw much attention."

The cannon dial is not beyond the reach of anyone. The small model in the Ernst Collection, see illustration opposite page 245, is only 4″ in diameter. Yes, it works, too.

XV

HUNTING SUNDIALS

W<small>E HAVE</small> shown you how to make dials. But suppose you do not want to do so? What then? Why not search for some dials to photograph? Or, for that matter, just to look at? See how many different kinds you can find; how many different designs; how many ways they function. Hunting for sundials can make any trip an adventure and much more interesting. It can take you off the beaten track and lead you to the discovery of many places of fascination and charm.

Some dials will tell only apparent time; some will have added furniture, such as lines of declination, an analemma, table for correcting the dial to standard time, or a linear graph to do the same thing; and some will have the whole correction incorporated into the design of the dial.

If you want to go hunting sundials, the locations of a few that are easily accessible may be helpful at the start. On the grounds of the Connecticut General Life Insurance Company headquarters in Bloomfield, Connecticut, is an 8-ton granite cube four feet on a side (pictured facing page 142). Its south face enables you to tell standard time noon, and a correction table on the north face provides the means to obtain standard time throughout each day of the year. The east and west faces also indicate the times of sunrise and sunset. A similar dial was erected by that company in Union Bank Square, Los Angeles, California.

Another interesting dial is on the cupola of the Bulova School of Watchmaking, 62nd Street, Woodside, Long Island, New York. It is a painted dial, about 3′ by 4′, and shows standard time noon. In Ottawa, Ontario, there are dials about 4′ by 6′ on the Mint Tower; and in Trois Rivieres, Quebec, there is a nice dial about 3′ by 5′ on the church of Des Ursulines. In Foxboro, Massachusetts, there is a large dial about 30′ by 40′ on the wall of the research building at the Foxboro Company (depicted opposite page 63). There is also a beautiful large dial incised in the wall of the Singing Tower, Lake Wales, Florida.

Many dials, both modern and old, will be found throughout Europe—look for them especially on churches and in parks.

If you want to photograph the dials that you find, use the best equipment you have. Of course, any camera will do, even the simplest Kodak or Polaroid, but your skill will largely determine the results.

Today the majority of pictures are taken in color, although we carry an extra camera that can be loaded with black-and-white film. In the following discussion, we shall emphasize color photography; some of our results are reproduced in the frontispiece and elsewhere throughout this book. The pictures were taken on Agfachrome, which has an ASA rating of 50, though Kodachrome is also good. Whatever your preference in color film, that is the film to use, because you know how to work with it.

For cameras, we have two M3 Leicas with f/2.8, 50-mm. lenses, each fitted with a sky filter. We also use a 135-mm. telephoto lens. In addition, we carry a Minox camera. Many dials are high up on a wall, and the telephoto makes it possible to fill the frame and show the details of the dial.

All of the pictures reproduced in color were taken in available light, without any fill-in flash. A Leica meter fitted to the cameras was used in all cases. You do not have to worry much about depth of field, unless you are taking pictures of dials at ground level, where close shots can be made without a telephoto.

What we like to do is take a picture of the whole scene, showing the position of the dial in relation to its surroundings. At ground level it is good to show a little bit of the setting of the dial—not too close nor too far away. The idea is to let the dial be the focal point and not be overpowered by the surrounding scenery.

After getting the pictures we want of the overall scene, we attach our telephoto, in the case of high dials, to bring out the details. For a dial at ground level, we move close up, to accomplish the same result. When using either lens we may take several pictures from various angles and distances. The selection of pictures in this book was made from five or six exposures of each dial.

The dial at Bowdoin College in Brunswick, Maine (shown opposite page 127), is a 3′ by 4′ block of limestone set into a brick wall. It is a prominent feature of the building and of pleasant color. Telephoto lens, f/8, 1/125 second.

An interesting dial, about 6′ by 8′ is on the Clausin Post Hotel in Garmisch, Germany. On the wall of the church in nearby Mittenwald (see picture opposite page 183) is a beautifully painted dial about 4' square. We photographed it at f/8, 125 second. Our particular choice of f-stop is often dictated by the color of the surrounding wall, especially if it contrasts sharply with the subject.

On the wall of a building in Market Square in Rothenburg, Germany (facing page 17), is a dial about 3' by 4', which we exposed at f/8, 1/125 second. In conjunction with the dial are a clock and two windows. At 11, 12, 13, and 14 hours the windows open and a brief pageant is performed to commemorate the drinking feat that saved the town from destruction in 1631. As the story goes, an enemy commander conquered the town and ordered it destroyed; but he offered a reprieve if one of the town councillors could empty a 3-liter mug of wine in a single draught. The old Mayor, Nusch, accomplished the feat, and the town was saved.

In Nordlingen, Germany, is another dial, about 8' by 10' on the side of the Red House. This white dial stands out in sharp contrast with the brilliant red facade of the building and is pictured opposite page 210. F/11 at 1/125 second.

On the wall of a home in Stamford, Connecticut, a yacht builder has worked his business into the scheme. As shown facing page 183, the gnomon was designed to be a prolongation of the ship's shadow. F/8 at 1/125 second.

The dial on the wall of a school in Stowe, Massachusetts, is in an ideal location. It can also serve for instructional purposes. See the large photograph opposite page 130. F/8 at 1/125 second.

When the Old State House on State Street in Boston, Massachusetts, was restored, it was decided to replace the clock with a sundial similar to the one that was originally

on the building. Old engravings showed the dial, so recon-
struction was possible (see facing page 210). It is about 4′
by 6′ and painted. F/5.6 at 1/125 second.

The foregoing dials are designed to tell apparent time
and are declining types. Now let us look at a few different
ones. One day we received a commission to design a dial
for a man who lived near the seashore, and wanted to put
a dial on the side of his garage that faced the sea, so
that his grandchildren could have something to indicate
the time when they were out sailing. This dial in Westport,
Connecticut, is 10′ by 12′ and is shown opposite page 44.
It shows the days of the year and an accompanying table
enables one to obtain standard time. Note the gnomon, the
base being in the form of a sextant. Although this is also
a declining dial, it has many different features. F/8, 1/125
second.

Atop Kitt Peak outside Tucson, Arizona, we designed an
equatorial dial (shown facing page 126) that stands at the
entrance to the museum. The design reflects the shape of
one of the telescopes nearby. The base is similar to that of
the telescope, the dial face is curved like a telescope tube,
and it is held in a fork, a common type of telescope mounting. Made of bronze, the dial is two feet across. It shows
the time of sunrise and sunset, the day of the year, and a
correction table enables one to find standard time. F/8,
1/125 second.

One of our favorite dials is the memorial to Dr. Lyman
Briggs, located on the grounds of the National Bureau of
Standards (depicted facing page 101). It is a polar dial and
is made of bronze. There are actually two dials, the primary
one on a flat plate 18″ by 24″ and bearing three gnomons.

Two of them indicate standard time throughout the year without correction by splitting the analemma into two segments; between these, the other gnomon indicates apparent time. At the top of the plate is a sunburst, pierced to allow a ray of light to fall on an arc at the back of the dial plate. The position of the ray or spot of light on the arc indicates standard time noon, the day of the year, and the altitude and declination of the sun. F/8, 1/125 second.

These pictures, as well as the others in this book, will give you some idea as to what kinds of dials you may find if you hunt for them. All present a different story. We have found dials on banks, churches, ordinary buildings, town halls, schools, private homes, and as memorials in parks. Why not hunt sundials—and don't forget your camera.

APPENDIX I

FORMULAS

T HE foregoing chapters show a simple, practical, and accurate method of laying out the hour lines for sundials, by means of a compass, protractor, and straight-edge. Sometimes it is desirable to compute the position of the hour lines, and lay them out on the dial plate by means of scales such as those used by engineers in plotting their surveys and other plans. Anyone familiar with trigonometry will not find the work of computing the hour lines arduous. An extensive knowledge of mathematics and astronomy is not necessary to fully understand the computations and formulas. Standard works on trigonometry and celestial mechanics will be indispensable to those who wish to delve into the theory of dialing and the derivation of the formulas discussed in this chapter.

The method of laying out the hour lines after their positions have been computed is fully described. Reference to the corresponding diagrams in the chapter on construction will aid the computer who is figuring a dial for the first time.

The desired hour lines may be plotted with a protractor; but a greater degree of accuracy will be obtained if the tangent method is used. By the tangent method, the numerical value for the tangent of the hour line angle is set off at right angles to the substyle (when the hour line angle is greater than 45° the cotangent is generally set off at right angles to the 6 p.m. line).

The formulas given here are those necessary for computing the position of the hour lines on the dials previously described. It must be remembered that the sun's hour angle is reckoned to the east or west from apparent noon, at which time the sun is on the meridian of the place; that 1h = 15°, 2h = 30°, etc.; and that the center of a dial is that point at which the hour lines converge.

TABLE SHOWING SUN'S HOUR ANGLE FROM 4 A.M. TO 8 P.M.
(APPARENT TIME)

Hour (Apparent Time)	Sun's Hour Angle (In Degrees)	Hour (Apparent Time)
12:00 noon	0° 00′	12:00 noon
11:30 a.m.	7° 30′	12:30 p.m.
11:00	15° 00′	1:00
10:30	22° 30′	1:30
10:00	30° 00′	2:00
9:30	37° 30′	2:30
9:00	45° 00′	3:00
8:30	52° 30′	3:30
8:00	60° 00′	4:00
7:30	67° 30′	4:30
7:00	75° 00′	5:00
6:30	82° 30′	5:30
6:00	90° 00′	6:00
5:30	97° 30′	6:30
5:00	105° 00′	7:00
4:30	112° 30′	7:30
4:00	120° 00′	8:00

The Equatorial Dial
Plate I

No formula is necessary for determining the position of the hour lines on this dial, since they are drawn from the center of the dial at regular intervals of 15° each.

The Horizontal Dial
Plate II

If we let

X = angular distance of hour lines from the substyle
L = latitude of the place
h = sun's hour angle in degrees
Then—

$$\tan X = \sin L \tan h$$

The Vertical South and North Dials
Plate III & IV

If we let

X = angular distance of hour lines from the substyle
L = latitude of the place
h = sun's hour angle in degrees
Then—

$$\tan X = \cos L \tan h$$

The Direct East—West Vertical Dials and the Polar Dial
Plate V & VI

Linear dimensions are used to compute the position of the hour lines on these dials instead of angular measurements, because they are parallel to each other and to the substyle.

Thus, if we let

X = linear distance of hour lines from the substyle
HS = height of the style in inches or millimeters
h = sun's hour angle in degrees

Then—

X = HS cot h (east and west vertical dials)
X = HS tan h (Polar Dial)

THE VERTICAL DECLINING DIALS
PLATE VII

By virtue of their position (vertical and declining), the computation of the hour lines for these dials is more complicated than for the foregoing. The distance of the substyle from the 12 o'clock line must be computed; the style's height above the dial plate must be determined; and in addition, the difference in longitude (difference between the meridian of the place and the meridian of the dial) must be found. When these facts are known, the position of the hour lines may be calculated.

Let

X = angular distance of the hour lines from the substyle
L = latitude of the place
h = sun's hour angle in degrees
SD = substyle distance from the meridian or 12 o'clock line
D = declination of the plane of the dial
SH = style's height
DL = difference in longitude

Then, for the substyle distance:

(1) tan SD = sin D cot L

For the height of the style:

(2) sin SH = cos D cos L

For the difference in longitude:

$$(3) \ \tan DL = \frac{\tan D}{\sin L} \ (\text{or}) \ \cot DL = \cot D \sin L$$

And, For the angular distance of the hour lines from the substyle:

(4) tan X = sin SH tan (DL ± h)

To avoid a false interpretation of formula 4, the computation of the hour lines for the dial shown in Plate VII is given below. This dial is a south vertical, declining 28°W in latitude 40°30′N.

Given:	L = 40° 30′	log sin SH = 9.82698 (Formula 2) = 42° 10′
	D = 28°	log tan DL = 9.91313 (Formula 3) = 39° 18′
	SD = 28° 48′ (Formula 1)	

Hours	h	DL + h	log tan DL + h	log sin SH	log tan X	X	
12 m	0°	39° 18′	9.91313	9.82698	9.74011	28°	48′*
11 a.m.	15°	54 18	0.14353	9.82698	9.97051	43	3
10	30°	69 18	0.42266	9.82698	0.24964	60	38
9	45°	84 18	1.00081	9.82698	0.82779	81	33
8	60°	99 18	0.78580	9.82698	0.61278	103	42
		DL − h	log tan DL − h				
1 p.m.	15°	24 18	9.65467	9.82698	9.48165	16	52
2	30°	9 18	9.21420	9.82698	9.04118	6	16
		SUBSTYLE falls between the hours of 2 and 3					
3	45°	5 42	8.99919	9.82698	8.82617	3	50
4	60°	20 42	9.57734	9.82698	9.40432	14	14
5	75°	35 42	9.85647	9.82698	9.68345	25	45
6	90°	50 42	0.08699	9.82698	9.91397	39	22
7	105°	65 42	0.34533	9.82698	0.17231	56	5

* This value is equal to the substyle's distance.

Note: Since this dial declines west, the substyle distance is measured to the right (east) of the 12 o'clock line; if it declined east, the substyle distance would be measured to the left (west) of the 12 o'clock line, and then the morning hours would be represented in formula 4 by DL—h and the afternoon hours by DL + h. For a further discussion of these dials, see pages 112-117.

THE DIRECT SOUTH AND NORTH RECLINING DIALS

These dials may be reduced to new latitudes, where they will become horizontal dials. With this new latitude known, the formula is the same as that for the horizontal dial.

Let

X = angular distance of the hour lines from the substyle
L = latitude of the place
h = sun's hour angle
R = reclination of the dial
Z = the new latitude where the reclining dial becomes a horizontal dial. (Z will also be the height of the style).

In the case of the south reclining dial:

If R $<$ (90° — L),   Z = (90° — L) — R
If R = (90° — L),   the dial is a polar dial, see page 110.
If R $>$ (90° — L),   Z = R — (90° — L)

In the case of the north reclining dial:

If R $<$ L,   Z = (90° — L) + R
If R = L,   the dial is an equatorial dial, see page 96.
If R $>$ L,   Z = 180° — [R + (90° — L)]

THEN:

$$\tan X = \sin Z \tan h$$

The Direct East and West Reclining Dials
Plate VIII

It will be readily seen that these dials can be reduced to those latitudes where they become vertical declining dials. This reduction is easily accomplished by the following formulas:

Let

Z = the new latitude where the dial will be vertical
D = the declination of the dial in the new latitude
R = reclination of the dial
L = Latitude of the place

Then:

$$Z = 90° - L$$

and

$$D = 90° - R$$

When these facts are determined, use the formulas for computing the position of hour lines for declining dials.

Thus, by substitution, we can derive formulas directly applicable to the direct east and west reclining dials.

If we let

X = angular distance of the hour lines from the substyle
L = latitude of the place
h = sun's hour angle in degrees
SD = substyle distance from the meridian or 12 o'clock line
R = reclination of the dial
SH = style's height
DL = difference in longitude

Then, For the substyle distance

$$(1) \quad \tan SD = \cos R \tan L$$

For the height of the style

(2) sin SH = sin L sin R

For the difference in longitude

(3) $\tan DL = \dfrac{\cot R}{\cos L}$ (or) cot DL = cos L tan R

AND, For the angular distance of the hour lines from the substyle:

(4) tan X = sin SH tan (DL ± h)

The application of formula 4 is the same as that for the declining dials. For clarity, the calculation of a direct east dial reclining 35° in latitude 39° N (see Plate VIII) is shown in the following table.

Given:	L = 39° 00'		log sin SH = 9.55746 (Formula 2) = 21° 10'			
	R = 35° 00'		log tan DL = 0.26427 (Formula 3) = 61° 27'			
	SD = 33° 33' (Formula 1)					

Hours	h	DL − h	log tan DL − h	log sin SH	log tan X	X
12 m	0°	61° 27'	0.26427	9.55746	9.82173	33° 33'*
11 a.m.	15°	46 27	0.02199	9.55746	9.57945	20 48
10	30°	31 27	9.78647	9.55746	9.34393	12 27
9	45°	16 27	9.47021	9.55746	9.02767	6 5
8	60°	1 27	8.40334	9.55746	7.96080	0 31
		SUBSTYLE falls between the hours of 7 and 8				
7	75°	13 33	9.38202	9.55746	8.93948	4 58
6	90°	28 33	9.73567	9.55746	9.29313	11 7
5	105°	43 33	9.97801	9.55746	9.53547	18 56
4	120°	58 33	0.21353	9.55746	9.77099	30 33
		DL + h	log tan DL + h			
1 p.m.	15°	76 27	0.61798	9.55746	0.17544	56 16

* This value is equal to the substyle's distance.
Note:—The hour lines as computed above will serve for a west dial having the same reclination in the same latitude; the morning hours would become the afternoon hours and the substyle would occupy a corresponding position among the afternoon hours; also the afternoon hours would be represented in formula 4 by DL − h. For a further discussion of these dials see page 117.

The Armillary
Plate X

The hour lines for this dial are laid out on the inner surface of that circle of the sphere representing the equator. Sometimes it is desirable to lay off the hour divisions by means of a tape, or millimeter flexible scale. The linear distance of the hour lines from the substyle may be computed from the following formula:

Where

h = sun's hour angle in degrees

R = radius of circle

X = linear distance of hour lines from the substyle, *measured on the dial plate*

C = value of h obtained from a table of circular arcs to Radius 1

Then:

$$X = CR$$

This formula needs no explanation, for it is solved by simple multiplication.

Another method of laying out the hour lines is shown on page 127 and notes concerning the armillary as a sundial will be found on page 129.

How to Compute the Azimuth

The azimuth of the sun must be found for dials which show the hour by the direction of a shadow cast by a perpendicular pin.

The formula for computing the azimuth is as follows,

where

D = Declination of the sun.
h = Sun's hour angle in degrees.
L = Latitude of the place.
A = Azimuth measured east or west from the south.

$$\text{Tan } N = \frac{\tan D}{\cos h}$$

Then

$$\text{Tan } A = \frac{\tan h \cos N}{\sin (L - N)}$$

How to Compute the Altitude of the Sun

The altitude must be found for each hour of the day when constructing pillar dials and others like them. First determine the angles N and A in the preceding formulas used to find the azimuth.

THEN:

$$\text{Tan Alt} = \cot (L - N) \cos A$$

Sunrise and Sunset

Using the notations given above, the azimuth of the rising or setting sun is determined by the following formulas:

$$\cos A = \sin D \sec L$$

or

$$\cos A = \frac{\sin D}{\cos L}$$

The hour angle, or time, of sunrise or sunset may be obtained from the following formula:

$$\cos h = \tan L \tan D$$

An hour angle expressed in degrees can be converted to time by using the table on page 245.

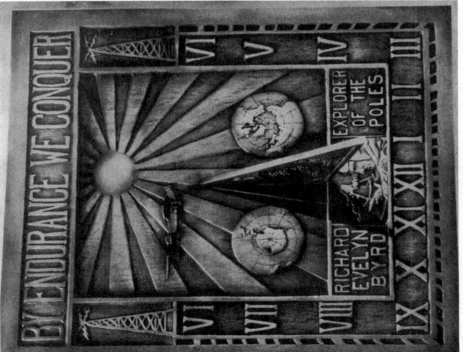

The Cathedral Landmark, Washington, D.C.

Exploration.

Signal Gun in Hamilton Watch Co. Collection, Lancaster, Penn.

Signal Gun in Ernst Collection, Harvard College Observatory.

Appendix II

TABLE 50.
TABLE FOR CONVERSION OF ARC AND TIME.

°	h. m.	°	h. m.	°	h. m.	' "	m s.	' "	m s.	' "	m. s.	' "	m. s.	"	s.
0	0 0	60	4 0	120	8 0	0 00	0 0	15 00	1 0	30 00	2 0	45 00	3 0	0	0.00
1	0 4	61	4 4	121	8 4	15	0 1	15	1 1	15	2 1	15	3 1	1	0.07
2	0 8	62	4 8	122	8 8	30	0 2	30	1 2	30	2 2	30	3 2	2	0.13
3	0 12	63	4 12	123	8 12	45	0 3	45	1 3	45	2 3	45	3 3	3	0.20
4	0 16	64	4 16	124	8 16	1 00	0 4	16 00	1 4	31 00	2 4	46 00	3 4	4	0.27
5	0 20	65	4 20	125	8 20	15	0 5	15	1 5	15	2 5	15	3 5	5	0.33
6	0 24	66	4 24	126	8 24	30	0 6	30	1 6	30	2 6	30	3 6	6	0.40
7	0 28	67	4 28	127	8 28	45	0 7	45	1 7	45	2 7	45	3 7	7	0.47
8	0 32	68	4 32	128	8 32	2 00	0 8	17 00	1 8	32 00	2 8	47 00	3 8	8	0.53
9	0 36	69	4 36	129	8 36	15	0 9	15	1 9	15	2 9	15	3 9	9	0.60
10	0 40	70	4 40	130	8 40	30	0 10	30	1 10	30	2 10	30	3 10	10	0.67
11	0 44	71	4 44	131	8 44	45	0 11	45	1 11	45	2 11	45	3 11	11	0.73
12	0 48	72	4 48	132	8 48	3 00	0 12	18 00	1 12	33 00	2 12	48 00	3 12	12	0.80
13	0 52	73	4 52	133	8 52	15	0 13	15	1 13	15	2 13	15	3 13	13	0.87
14	0 56	74	4 56	134	8 56	30	0 14	30	1 14	30	2 14	30	3 14	14	0.93
15	1 0	75	5 0	135	9 0	45	0 15	45	1 15	45	2 15	45	3 15	15	1.00
16	1 4	76	5 4	136	9 4	4 00	0 16	19 00	1 16	34 00	2 16	49 00	3 16	16	1 07
17	1 8	77	5 8	137	9 8	15	0 17	15	1 17	15	2 17	15	3 17	17	1.13
18	1 12	78	5 12	138	9 12	30	0 18	30	1 18	30	2 18	30	3 18	18	1 20
19	1 16	79	5 16	139	9 16	45	0 19	45	1 19	45	2 19	45	3 19	19	1.27
20	1 20	80	5 20	140	9 20	5 00	0 20	20 00	1 20	35 00	2 20	50 00	3 20	20	1.33
21	1 24	81	5 24	141	9 24	15	0 21	15	1 21	15	2 21	15	3 21	21	1 40
22	1 28	82	5 28	142	9 28	30	0 22	30	1 22	30	2 22	30	3 22	22	1 47
23	1 32	83	5 32	143	9 32	45	0 23	45	1 23	45	2 23	45	3 23	23	1.53
24	1 36	84	5 36	144	9 36	6 00	0 24	21 00	1 24	36 00	2 24	51 00	3 24	24	1.60
25	1 40	85	5 40	145	9 40	15	0 25	15	1 25	15	2 25	15	3 25	25	1.67
26	1 44	86	5 44	146	9 44	30	0 26	30	1 26	30	2 26	30	3 26	26	1.73
27	1 48	87	5 48	147	9 48	45	0 27	45	1 27	45	2 27	45	3 27	27	1.80
28	1 52	88	5 52	148	9 52	7 00	0 28	22 00	1 28	37 00	2 28	52 00	3 28	28	1.87
29	1 56	89	5 56	149	9 56	15	0 29	15	1 29	15	2 29	15	3 29	29	1 93
30	2 0	90	6 0	150	10 0	30	0 30	30	1 30	30	2 30	30	3 30	30	2.00
31	2 4	91	6 4	151	10 4	45	0 31	45	1 31	45	2 31	45	3 31	31	2.07
32	2 8	92	6 8	152	10 8	8 00	0 32	23 00	1 32	38 00	2 32	53 00	3 32	32	2.13
33	2 12	93	6 12	153	10 12	15	0 33	15	1 33	15	2 33	15	3 33	33	2.20
34	2 16	94	6 16	154	10 16	30	0 34	30	1 34	30	2 34	30	3 34	34	2.27
35	2 20	95	6 20	155	10 20	45	0 35	45	1 35	45	2 35	45	3 35	35	2.33
36	2 24	96	6 24	156	10 24	9 00	0 36	24 00	1 36	39 00	2 36	54 00	3 36	36	2.40
37	2 28	97	6 28	157	10 28	15	0 37	15	1 37	15	2 37	15	3 37	37	2.47
38	2 32	98	6 32	158	10 32	30	0 38	30	1 38	30	2 38	30	3 38	38	2.53
39	2 36	99	6 36	159	10 36	45	0 39	45	1 39	45	2 39	45	3 39	39	2.60
40	2 40	100	6 40	160	10 40	10 00	0 40	25 00	1 40	40 00	2 40	55 00	3 40	40	2.67
41	2 44	101	6 44	161	10 44	15	0 41	15	1 41	15	2 41	15	3 41	41	2.73
42	2 48	102	6 48	162	10 48	30	0 42	30	1 42	30	2 42	30	3 42	42	2.80
43	2 52	103	6 52	163	10 52	45	0 43	45	1 43	45	2 43	45	8 43	43	2.87
44	2 56	104	6 56	164	10 56	11 00	0 44	26 00	1 44	41 00	2 44	56 00	3 44	44	2.93
45	3 0	105	7 0	165	11 0	15	0 45	15	1 45	15	2 45	15	3 45	45	3.00
46	3 4	106	7 4	166	11 4	30	0 46	30	1 46	30	2 46	30	3 46	46	3 07
47	3 8	107	7 8	167	11 8	45	0 47	45	1 47	45	2 47	45	3 47	47	3.13
48	3 12	108	7 12	168	11 12	12 00	0 48	27 00	1 48	42 00	2 48	57 00	3 48	48	3.20
49	3 16	109	7 16	169	11 16	15	0 49	15	1 49	15	2 49	15	3 49	49	3.27
50	3 20	110	7 20	170	11 20	30	0 50	30	1 50	30	2 50	30	3 50	50	3.33
51	3 24	111	7 24	171	11 24	45	0 51	45	1 51	45	2 51	45	3 51	51	3 40
52	3 28	112	7 28	172	11 28	13 00	0 52	28 00	1 52	43 00	2 52	58 00	3 52	52	3.47
53	3 32	113	7 32	173	11 32	15	0 53	15	1 53	15	2 53	15	3 53	53	3.53
54	3 36	114	7 36	174	11 36	30	0 54	30	1 54	30	2 54	30	3 54	54	3.60
55	3 40	115	7 40	175	11 40	45	0 55	45	1 55	45	2 55	45	3 55	55	3.67
56	3 44	116	7 44	176	11 44	14 00	0 56	29 00	1 56	44 00	2 56	59 00	3 56	56	3.73
57	3 48	117	7 48	177	11 48	15	0 57	15	1 57	15	2 57	15	3 57	57	3.80
58	3 52	118	7 52	178	11 52	30	0 58	30	1 58	30	2 58	30	3 58	58	3.87
59	3 56	119	7 56	179	11 56	45	0 59	45	1 59	45	2 59	45	3 59	59	3.93
60	4 0	120	8 0	180	12 0	15 00	1 0	30 00	2 0	45 00	3 0	60 00	4 0	60	4 00

DEGS/15= H,m
MIN-SEC/15=M.S.

INDEX